THE
ELMUSTEE
TECHNOLOGIES

A.H. CEMENDTAUR

ISBN-10: 1523333952
ISBN-13: 978-1523333950
US$33.35

To the last person who will be forced to leave their ancestral home because of an ongoing political or economic turmoil in their land

CONTENTS

Preface

Elmustee, a membership community, is determined to be self-sustaining in food, water, and energy. Elmustee uses technologies that help the community meet its sustenance standards at a minimum cost to the Elmustee members.

Besides providing security--security being a member's very basic expectation from a social structure--the latest technologies used at Elmustee enable the community to meet its social welfare goal of providing affordable food, water, shelter, education, and healthcare to all community members.

This book describes the technologies and strategies being used at Elmustee to meet the food, water, energy, and other needs of the community members.

Ali Hasan Cemendtaur

August 2, 2016

1 Member induction

Synopsis

Elmustee members go through an application process that helps applicants orient themselves with the philosophy of the membership community they wish to become a part of.

Procedure

Applicants fill out a simple online form providing their identifying information (name, gender, date of birth, place of birth, height, weight, mother's name, physical address, telephone number, and an email address). This information is needed to uniquely identify each member.

Each applicant also provides names and membership numbers of two current Elmustee members the applicant personally knows. This information helps Elmustee administrators reach the potential member even when that person cannot be directly contacted.

On completion of the online application form and paying the application fee, the potential member is issued an applicant member. The applicant uses this number to log in at the Elmustee web site and watch orientation videos present there. The short and entertaining videos on topics as varied as social contract, food, water, and shelter (see the book 'Elmustee, The Membership Experience') help the applicant understand the philosophy behind Elmustee, and prepare them for a life in the citizenship-focused community. The web site keeps track of the time the applicant spends in watching the orientation videos.

Once the applicant has watched all the videos, the applicant is ready to appear for the literacy testing at the Elmustee office. Elmustee expects its members to be literate, for them to follow written instructions and to communicate in writing with other community members.

The literacy test is given on a computer at the Elmustee office. The test can be taken in three languages: in English, in Urdu, and in Sindhi. There is no fee if the test is taken in Sindhi, and has a discount when taken in Urdu; standard charges apply for the English test.

At the testing center, on confirmation of the applicant's identity, the applicant is directed to a station where on a computer screen the applicant sees the three options in text form, of specific language they wish to take the test in. On pressing the onscreen button of the selected language the test starts in that language.

First, the applicant is tested on their reading skills. A paragraph of normally written text appears. Below the passage there are questions related to the text. Applicant reads the text and the questions, and then answers every question using an answer from multiple choices appearing on the computer screen. There are two passages; each passage is followed by three questions.

After being tested on the reading skills, the applicant takes a writing skill test.

In the writing skill test members are asked to use the onscreen keyboard to answer the questions appearing on the screen. This test too is comprised six questions.

After the completion of the two tests, the applicant is notified if they passed the literacy test.

If the applicant passes the literacy test, they are randomly tested on the material provided in the orientation videos.

On passing the computerized tests the member is ready for biometrics. A biometrics-booth (see 'Membership, Identification, and border control' section) is used to weigh the member, measure their height, their torso, and take their fingerprints.

A membership ID card bearing the photo and other identifiable information is issued to the member. Each member is given a unique membership number. The color (green or red) of the ID card indicates if the member is a resident or a nonresident. All members are initially issued a red (nonresident) ID card. Only on establishing residence in Elmustee, a member's nonresident card is replaced with a green (resident) ID card. The red ID card allows a member to visit all areas of the Elmustee Community except for the residents-only area (a small area comprising permanent housing). The large area, in Elmustee, open to all members includes the library, the community hall, pool, hiking tracks, the campground, parks, etc.

Elmustee School

Because it is an individual's choice to become a part of our community, it is an individual's call if they are willing to develop the basic skills that would enable them to get into our community. But Elmustee goes beyond providing a membership community to the qualified denizens of the land. Elmustee is ready to *make* people qualify for the community's membership. Elmustee is ready to help everyone and anyone. And it is ready to help even those who have nothing else but a desire to become a part of our community.

Besides Elmustee's City Testing Center, Elmustee runs a school where individuals--including those who do not even know how to read and write--are trained to become members of our community. The school runs on donations given by people interested in making individuals qualify for admission to our community. The Elmustee School prepares students in basic literacy; individuals are then made to go through the course material that comprises the videos everyone interested in becoming a part of our community is asked to watch.

Conclusion & Specific Comments

Orientation videos made by Elmustee can be used by people and institutions agreeing to Elmustee's philosophy of training people. These videos describe what members should expect to see inside Elmustee, including the water, electricity generation, and other systems being used in the Community. Elmustee's membership criteria is simple to follow: 1. Provide basic information about yourself, 2. Provide two references of current Elmustee members, 3. Pay the application fee, 4. Take the literacy test, 5. Take a test on Elmustee orientation videos, and 6. Go through the biometrics collection process.

Elmustee Member Induction

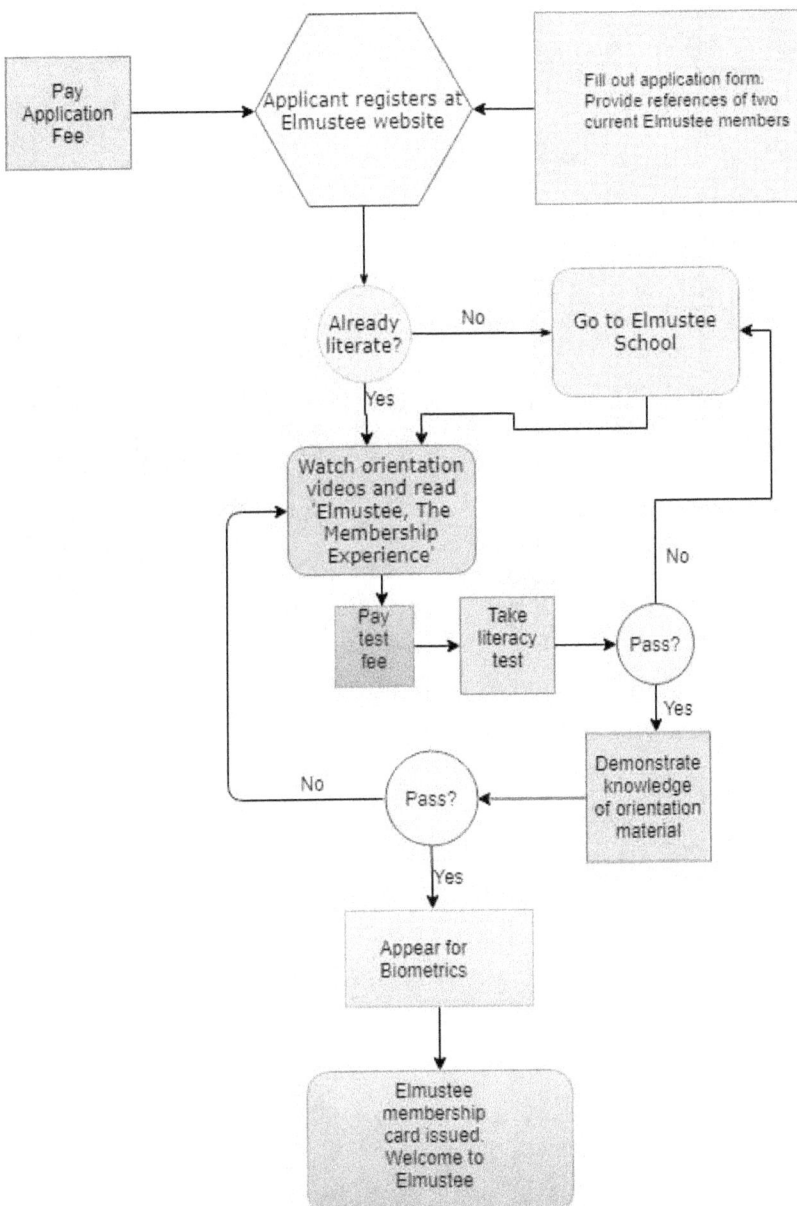

Figure 1 Elmustee member induction process

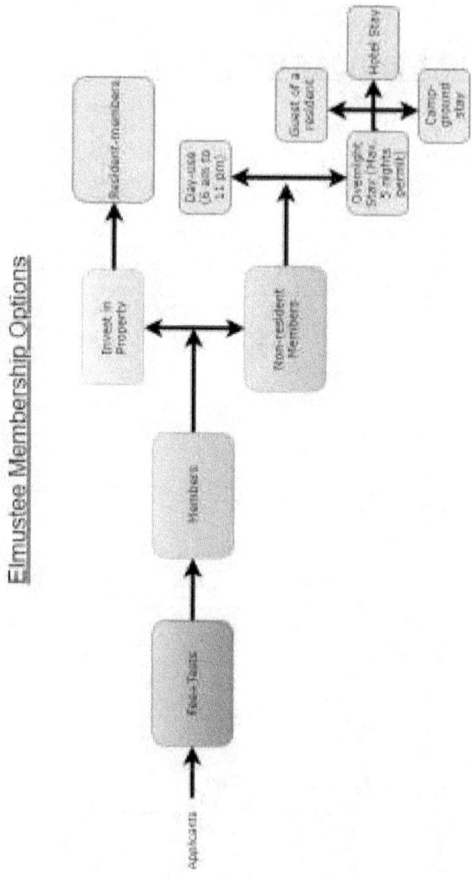

Figure 2 Elmustee membership options

2 General security

Synopsis

Elmustee is a membership community. In order to provide a safe environment to its members, Elmustee secures its geographical boundary. Multiple layers of security measures (a moat, a boundary wall, an electronic intrusion-detection system, surveillance cameras, multiple security posts, and surveillance drones) are used at Elmustee.

Description

- State of the Art

Physical barriers, electronic surveillance and intrusion detection technologies aid Elmustee Security Company in fighting intrusion attempts.

- Technology

Elmustee's first line of defense against intrusion is its fleet of surveillance drones. Regularly flown from Elmustee, each drone, fitted with high resolution cameras, scans Elmustee environment to detect any possible intrusion.

Elmustee's first *physical* barrier against intrusion is the moat surrounding the community. This moat, over 7 feet deep, is filled with seawater.

Approaching the community, the moat is followed by an electronic intrusion detection system. Laser posts and receptacles are installed a few feet away from the moat to detect intrusion. This detection system notifies the Elmustee Security Company of an intrusion attempt. The Elmustee Security Company is responsible for manning the security posts. Once set off, the intrusion detection system sounds alarms, turns on blinding lights, ignites firecrackers, and plays recorded warning messages to deter intruder(s) from advancing any further.

A six feet boundary wall dotted with security posts is our ultimate combat-ready defense against illegal movement across the Elmustee boundary. When needed armed guards rush to the identified area and take positions at the security posts.

- Application

After earth-moving equipment moved the dirt out of the ditch around the Elmustee Community, the ditch was lined with three layers of impermeable membrane. Moisture detection sensors are present between the first and the second, the second and the third membranes, and under the third membrane. Since seawater is used to fill up the moat, this leak detection system is an important feature of this waterbody.

First Cost

Costs of making the necessary components of the Elmustee General Security system including the cost of making a moat; of installing pumps to fill up the moat; of a fleet of drones; of an electronic Intrusion Detection System (EIDS); of an audio system, and its integration with the EIDS; of the fire-crackers, and their integration with the EIDS; of the boundary wall; of the security posts; of guns, etc. are provided in a separate spreadsheet.

The contractor winning the contract for the operation of the general security system for the first three years will build the security system using a loan from Elmustee.

Operating Costs

Elmustee awards an annual contract to a successful bidder for the maintenance and operation of the Elmustee General Security System. The contractor is responsible for providing manpower to keep the moat operational, for running the water pumps, for regular dredging, for the operation of the drones, for the upkeep of the electronic intrusion detection system and the associated audio system and the firecrackers, for manning the posts, for providing the ammunition, etc.

Conclusion & Specific Comments

It is impossible to build an intrusion deterrent system that is 100% effective, but using the latest technologies Elmustee is building a security system that will win the confidence of the Elmustee members. Using modern technologies Elmustee is reducing the operating cost of the security system.

Figure 3 Elmustee uses drones to keep an eye on its surroundings

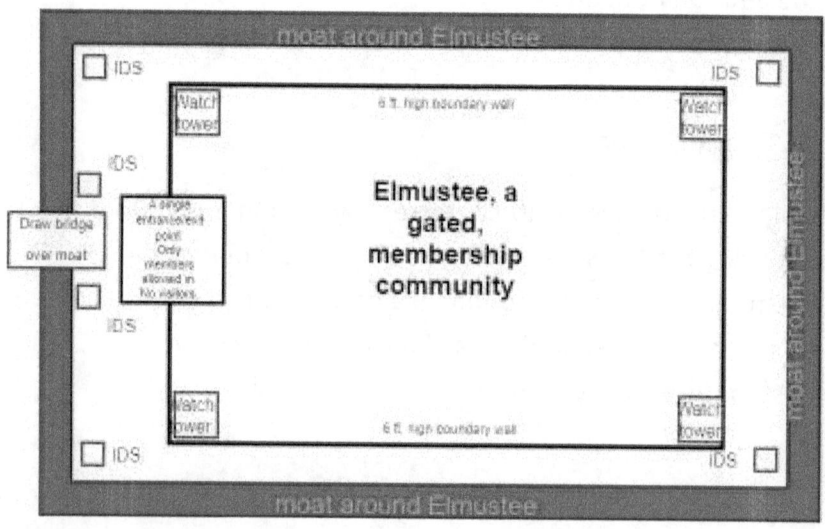

Figure 4 Elmustee security layout

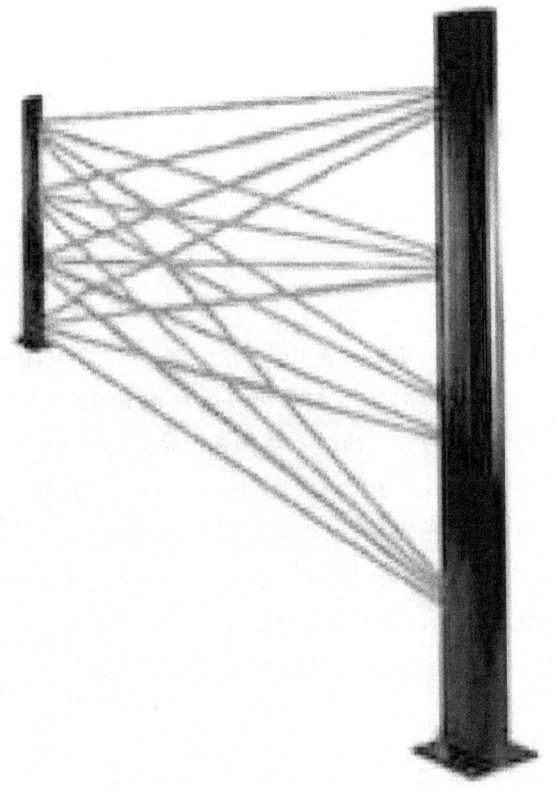

Figure 5 Intrusion detection system

3 Membership, identification, and border control

Synopsis

Elmustee strives to be a crime free society. It is imperative that Elmustee members and administrators know the identity of people physically present on community's premises. General Elmustee membership allows members to have access to all Elmustee areas except the residents-only areas. Elmustee members are issued integrated-circuit ID cards. The ID card is used to enter the general Elmustee area, and, for resident-members, to enter the residents-only areas of Elmustee.

Description

- State of the Art

On completing the membership process a photo ID is issued to the member. There are distinct (green and red) ID cards for resident and nonresident members. The ID card is renewed annually on the payment of the annual membership dues and on completion of the required continuing education credits.

Members park their vehicles in the parking area on the country side of the moat. They cross the bridge over the moat through a narrow passage that allows room for a one-person line. On entering the reception area members pay the entrance fee and enter the automated identification booths. Each booth can accommodate only one person at a time. On confirming the identity of the member, a door leading to the entrance of Elmustee is opened and the member enters the community.

A large part of the Elmustee community is accessible to all members, residents and non-residents alike. The all-members-areas include the campground, hotels, the community center, the commercial center, medical facilities, many parks, libraries, office buildings, hiking tracks, etc. A small physical portion of the community is only open to the resident/renters of the community—residences, schools, and some parks are part of that restricted area.

Non-residents' stay in Elmustee

There are several ways non-resident Elmustee members can stay at Elmustee for a longer period of time. They can stay with an Elmustee resident, or can use any of the hotels, or the campground facility. Guests staying at a private residence need to obtain an overnight-guest ID from the reception—service charges apply. The overnight-guest ID will allow the guest access to the residents-only-restricted area.

Elmustee ID cards

Elmustee issues two types of ID cards to its members. Resident members are issued a green ID card; non-resident members get red ID cards; overnight-guests staying with an Elmustee resident get temporary (per night) authorization to enter the residents-only area of the community. All ID cards are scan-able.

Note on Honorary Membership

Elmustee administrators understand the importance of keeping the movers and shakers of the larger society by its side. On a case by case basis, honorary membership is given to key personnel in federal and local governments, in the army, the police, and to top leaders of major political parties.

Elmustee actively facilitates membership of law enforcement personnel serving the area Elmustee is located in. This ensures that the Elmustee members-only entrance criterion will not be compromised if law enforcement personnel need to officially visit the community.

- Technology

The ID booths inside the reception hall are equipped with technologies that confirm the identity of a member through multiple ways. As soon as a member scans their ID card, the data related to the particular member becomes available at the booth's computer.

When in the booth, the member is asked to put all bags and backpacks in a bin provided for such storage. The member stands on a platform, putting their hands on touchpads.

Member is weighed and the current weight of the person is compared with the historical data. Through light sources on one wall and interceptors on the other, member's height, and girths around the shoulder and torso are measured and compared with the historical data. The member in the booth is asked to put their hands on touchpads; their fingerprints too are matched with the historical data. When the new data matches with the historical biometric data to a large extent, to an accuracy of 90%, the member is allowed entrance into Elmustee. When the two sets of data do not match, the member is asked to go for a manual admission process--there is an extra charge to administer the manual admission process.

Male to female ratio in Elmustee

Because of the self-sustaining nature of Elmustee, the community allows only a fixed number of members on the premises. Currently that number is 10,000. This number includes our 5000 resident members. Consequently, only 5000 non-resident members are concurrently allowed inside. A counter at the moat informs arriving people about the latest number of non-resident members present inside--this number is also updated on Elmustee's web site. Since Elmustee has over 8000 non-resident members, the non-residing members are requested to check Elmustee web site for community's latest head count, and male to female ratio before making a trip to Elmustee. Even when the counter shows a full house, arriving members are requested to wait--there are always people leaving the premises.

In some cases even when the total population of Elmustee is below its maximum at a given time, male members are asked to wait at the gate—this happens when the admission of male members is calculated to decrease the female population of the community below 50%. Please note that Elmustee has decided to keep the female population of the Community at least 50%, at any given time. As an example, consider this scenario. It is 10 am on a Tuesday. The population counter at the Elmustee reception building is showing a current male population of 2385, and the female population is 2386. Two male members arrive at the reception. Only one of them can enter at this time because admission of both male members will decrease the female population to below 50%. The other male member has to wait till either another male member leaves the premises, or a female member of the community shows up for admission.

First Cost

Costs of making the necessary components of the Elmustee Border Control System (including the cost of making the 'Identification booths') are provided in a separate spreadsheet.

The contractor winning the contract for the operation of the Elmustee Security System for the first three years will build the Elmustee Border Control System using a loan from Elmustee.

Operating Costs

Elmustee awards an annual contract to a successful bidder for the maintenance and operation of the Elmustee Border Control System.

The contractor is responsible for running the operation of the Elmustee Border Control System, and for taking care of any needed repairs of the 'Identification booths', etc.

Benefits

Elmustee border control relies heavily on the latest biometric identification technologies.

Even though the admission process is cost effective, the continuous traffic through the Elmustee border puts a financial burden on the community. To cover costs, Elmustee charges an admission fee every time a member enters the community. The admission fee adds to Elmustee's revenue stream.

Since there is no charge on leaving the community--fee is collected on entrance only--Elmustee hopes to see manageable traffic across the Elmustee border; this reduced traffic will in turn make it easy to monitor the border and stop illegal admission.

Conclusion & Specific Comments

Strict border control at the Elmustee boundary helps to make Elmustee a safe place. Elmustee's day visitors add to the community's income stream.

Elmustee actively facilitates membership of law enforcement personnel serving the area the community is located in.

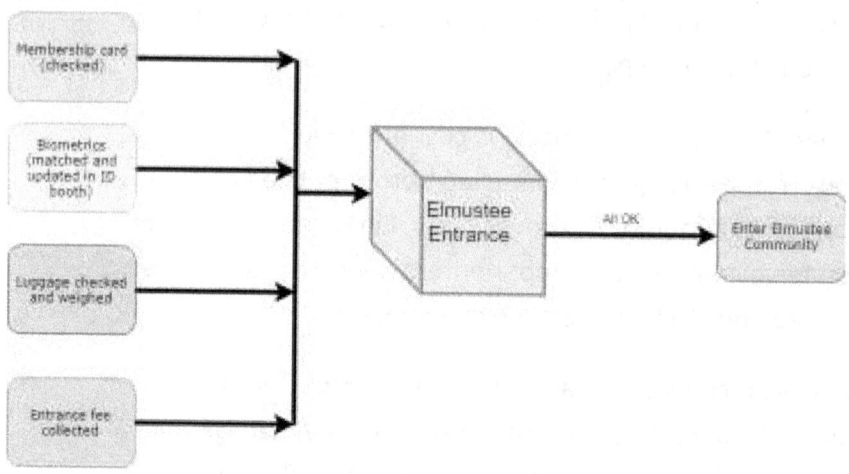

Figure 6 Elmustee entrance checks

Figure 7 ID booth at the entrance

4 Elmustee surveillance system

Synopsis

Elmustee strives to be a crime free society. All general areas in Elmustee are under 24-hour video and audio surveillance.

Description

- State of the Art

Surveillance cameras have become ubiquitous. Law enforcement agencies all over the world use video footage from surveillance cameras to solve crimes. Elmustee uses video cameras to record all activities in the general areas of the community. The video footage is not normally monitored live. The footage is stored, for years; usually it is retrieved only when a particular incident needs to be investigated.

Being under the surveillance of the cameras, you will not find privacy in the public areas of Elmustee. Elmustee members have agreed to sacrifice their privacy in the public areas in order to have a strong crime-deterrent atmosphere.

- Technology

Elmustee uses low and high resolution cameras, and audio-recorders to record all activities taking place in the general areas of the community. The captured video and audio streams are saved at different places for a duration of at least thirteen months. New footage is written over footage that is more than thirteen-month old.

- Application

The first set of low-resolution, wide angle cameras are installed under the photovoltaic (PV) panels, at a height of 30 ft.--each camera covers roughly a half-circle of 40 feet diameter. These cameras provide a bird's eye view of the general areas. Video and audio footages from these cameras are stored at a location known as Storage A. A second set of low resolution, wide angle cameras are installed at a height of 15 ft.--their field of vision is roughly 20 feet. Audio and video footages obtained from these cameras are stored at Storage B. Installed at a height of 5 ft. to 6.5 ft. are our high resolution, narrow angle cameras providing clear photos of individuals, and audio of conversations and actions; these cameras help us in identifying faces and in solving a crime. Video and audio footage from these high resolution cameras is stored at Storage C.

First Cost

Costs of making the necessary components of the Elmustee Surveillance System (including the cost of surveillance cameras and their installation; the cost of three different data storage systems) are provided in a separate spreadsheet.

The contractor winning the contract for the operation of the Elmustee Security System for the first three years will build the Elmustee Border Control System using a loan from Elmustee.

Operating Costs

Elmustee awards an annual contract to a successful bidder for the maintenance and operation of the Elmustee Surveillance System.

The contractor is responsible for smoothly operating the Elmustee Surveillance System, and for taking care of any needed repairs.

Benefits

Since Elmustee is a membership community and the administration office keeps a record of everyone present on the property, and because of the continuous video and audio recording of the premises, a strong deterrent for pre-meditated crime exists. Spontaneous crime cases, if any, are solved through video and audio footage stored at three distinct locations.

Conclusion & Specific Comments

Knowing the identity of people present inside Elmustee, and video and audio surveillance of the general areas of Elmustee help the community in maintaining its near crime free status.

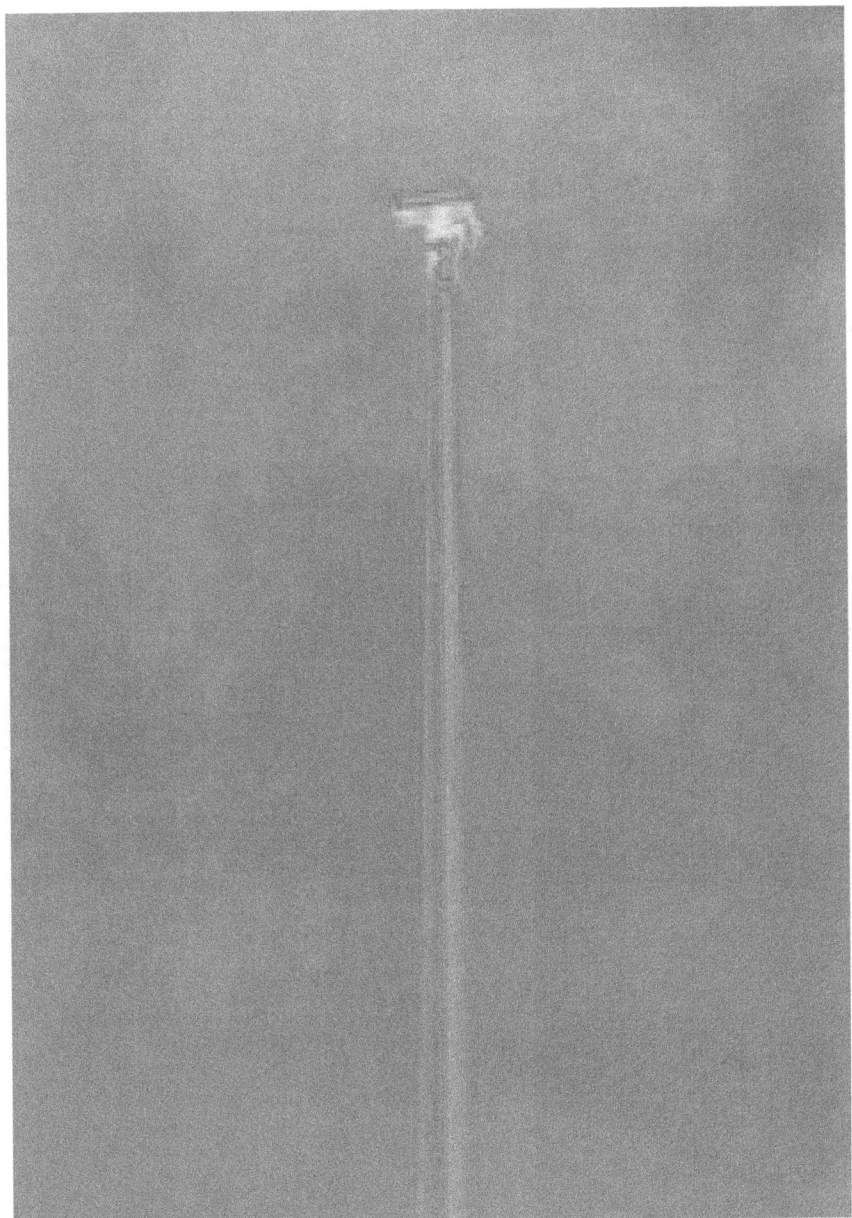

Figure 8 High level surveillance cameras

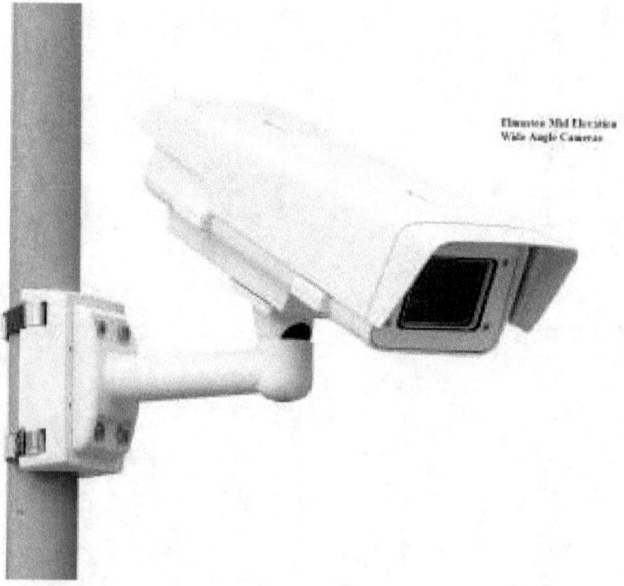

Figure 9 Mid-level surveillance cameras

Figure 10 Eye level cameras

5 The Elmustee financial system

Synopsis

Elmustee financial system ensures manageable inflation in prices in order to keep food, water, and other essentials affordable for the Elmustee members.

Description

- State of the Art

Elmustee uses its own currency--the currency is only available in electronic format. Elmustee members charge their ID cards with Elmustee currency, and use this currency as 'cash' when making a purchase inside Elmustee. Elmustee currency is pegged to a weighted average of strong international currencies (USD, Yen, Yuan, Euro, and GBP).

- Technology

Elmustee members use cash or checks from outside to charge their ID cards with the Elmustee currency. Members are able to see online all credit and debit charges made to their ID cards. Members use their ID cards as normal bank debit cards to make purchases inside Elmustee. Members can also use debit and credit cards from outside Elmustee--currency conversion charges apply.

- Application

Elmustee financial administrators set the Elmustee currency conversion rate based on the latest bank rates and the relative influence of particular economies.

First Cost

Cost of enabling the ID cards to store Elmustee currency data

Development of a web system for keeping track of Elmustee currency; and for people to check their transactions within Elmustee.

Operating Costs

Maintenance of the web-based currency system

Benefits

Electronic payment through member ID cards ensures hassle free financial transactions while making it easy for the Elmustee administration to collect sales tax.

Conclusion & Specific Comments

Having its own currency and the state-of-the-art electronic payment system help Elmustee maintain transparent financial records.

Incomes and Payments in Elmustee

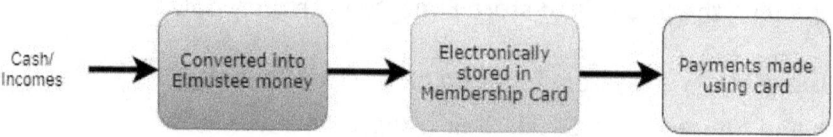

Figure 11 Incomes and payments in Elmustee

6 The Elmustee transportation system

Synopsis

The Elmustee transportation system is fueled mainly by electricity and animal power. Small-sized electric cars (golf carts) equipped with batteries charged by fuel cells and/or photovoltaic panels ply Elmustee roads. An electric railroad links Elmustee entrance to the Community Center and the Residential area.

Description

- State of the Art

Whereas there is a trend towards the use of electric vehicles, today's transportation still mainly relies on fossil fuels. With this heavy reliance on fossil fuels to run a very important sector of the world economic engine, the danger of total collapse of the economy in case of a failure related to the extraction or transportation of the fossil fuels is very much a possibility. Elmustee consciously uses only renewable energy sources to meets its transportation needs.

Most Elmustee members find it convenient to walk or bike from one place to another, within Elmustee.

- Technology

Motorized transportation

In Elmustee all motorized vehicles run on electricity. The large animal population of Elmustee--along with the human population--generates enough waste to produce adequate amount of methane gas. Toilet water from Elmustee buildings, and animal manure make their way to the biogas plants at Elmustee. These biogas plants generate methane that is then used in fuel cells, or is stored in compressed gas cylinders using animal power. Elmustee's photovoltaic array is used to charge electric batteries used by cars.

Elmustee's mass transit system is a railroad that has its one end just outside the Elmustee boundary (the station known as the 'Loading Station'), the second end at the Community Center, and the third end at Elmustee residences. Electric trains, running on underground cable, make frequent trips between the three stations. Power for the electric trains is generated through the fuel cells that in turn use the biogas from the biogas plants.

Walk-ways

Within Elmustee, walkways shaded with trees and vines are used by the Elmustee members to walk from one place to another.

Bike-paths

Two-way paved roads, only open to bicycles, connect various points of interest within Elmustee. Bike parking racks are present at various locations. Elmustee members can either bring their own bikes from outside, or buy or rent bicycles within Elmustee.

- Application

Elmustee driver licenses are issued to members after they pass a written and a driving test. A distinct ID card indicates a particular member's driving privilege in Elmustee.

Most Elmustee members either use sidewalks or bike paths to go from one place to another, within Elmustee.

Elmustee electric train, mostly used by the elderly and differently-abled people, runs between the three stations every 30 minutes.

Electric-powered taxis can be hired by Elmustee members.

A few Elmustee members own their own electric cars.

First Cost

First Cost related to the railroad and trains, walkways and bike-ways, etc. are being presented in a separate spreadsheet.

Operating Costs

Maintenance and operating costs of the Elmustee transportation system are being presented in a separate spreadsheet mentioned above.

Benefits

The Elmustee transportation system is designed to keep the Elmustee members healthy, the Elmustee environment free of pollution, and for the community to have zero reliance on outside energy resources.

Conclusion & Specific Comments

Since outside vehicles are not allowed inside Elmustee, the Community's traffic system is easy to manage. Because the Elmustee vehicles only use renewable energy, the Community does not depend on outside energy sources for its transportation needs.

Elmustee Energy for Transportation

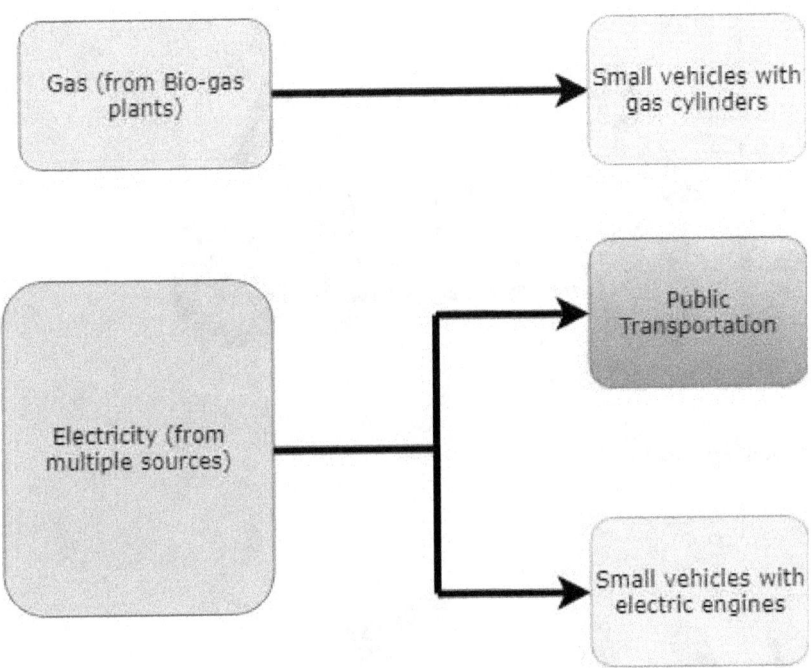

Figure 12 Elmustee transportation energy sources

Figure 13 Elmustee taxis

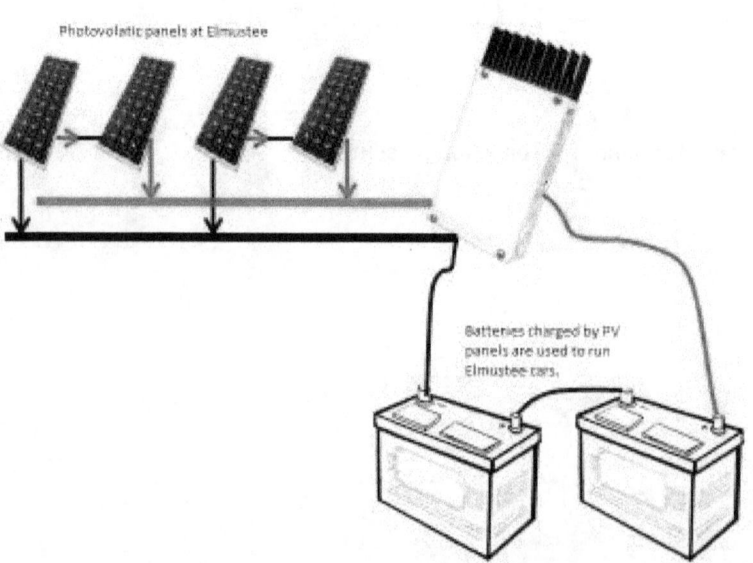

Figure 14 PV panels used to charge car batteries

7 Elmustee general construction

Synopsis

Construction in Elmustee is attuned to the geography the community is located in. The specialized construction suitable for extremely hot weather ensures that the air conditioning load of Elmustee buildings is greatly decreased.

Description

- State of the Art

Construction in Karachi uses reinforced concrete skeleton and roof, with concrete blocks making up the walls. Whereas flat roof surfaces are the biggest contributors of heat, roof Insulation is seldom used in these buildings. Consequently, buildings made in this fashion absorb and retain heat in the structure envelope; considerable amount of electrical energy is then used in meeting the high air conditioning load thus produced.

- Technology

In Elmustee all buildings are built under the shade of the photovoltaic array--see passive cooling section of this document for further elaboration of the concept.

Elmustee buildings are required to have insulated walls and roofs. Air-conditioning load of a building is met with fan-coil units present inside the building. Metered chilled water is provided to each building.

- Application

First the shade under which buildings are constructed, and then the envelope insulation of the buildings reduce heat entering the structures. Thus, for the most part the cooling load of the Elmustee buildings is the internal load--heat added by the people, lights, equipment--of the buildings. The absence of condensers (heat rejection equipment) at the buildings, further helps in keeping the ambient air around structures, cooler.

First Cost

Construction cost of the community buildings is presented in a separate spreadsheet.

Individual fan-coil and air-handling units are installed by the building owners.

See section on district-cooling for the cost of site chilled water system.

Operating Costs

Maintenance and operating costs of the community buildings are presented in a separate spreadsheet mentioned above.

Individual fan-coil and air-handling units are maintained by the building owners.

See section on district-cooling for the maintenance cost of site chilled water system.

Benefits

The specialized construction of Elmustee buildings ensures the maintenance of pleasant indoor temperature, at a minimum operating cost.

Conclusion & Specific Comments

Elmustee buildings rise to the challenge of construction appropriate for hot climate.

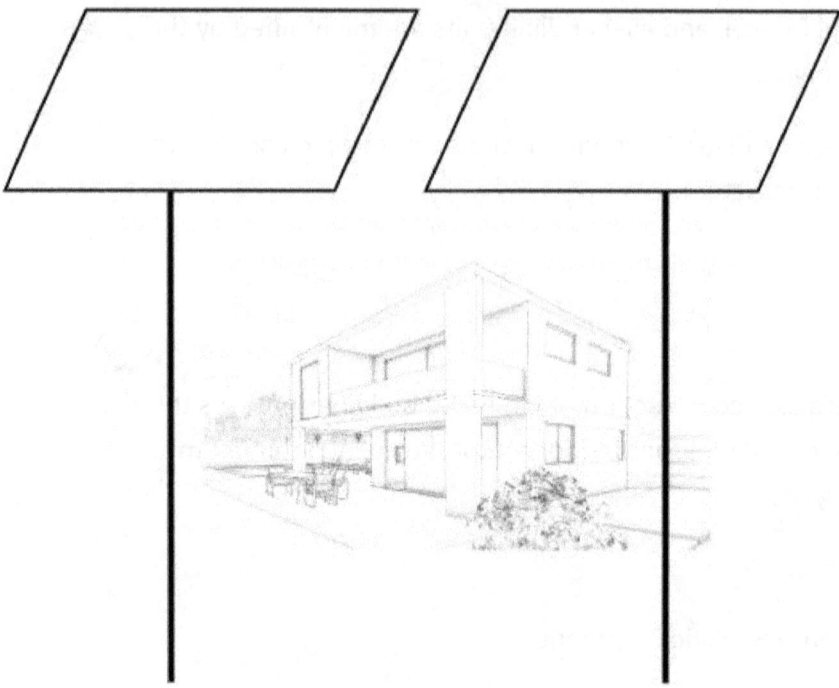

Figure 15 Elmustee houses built under PV panels

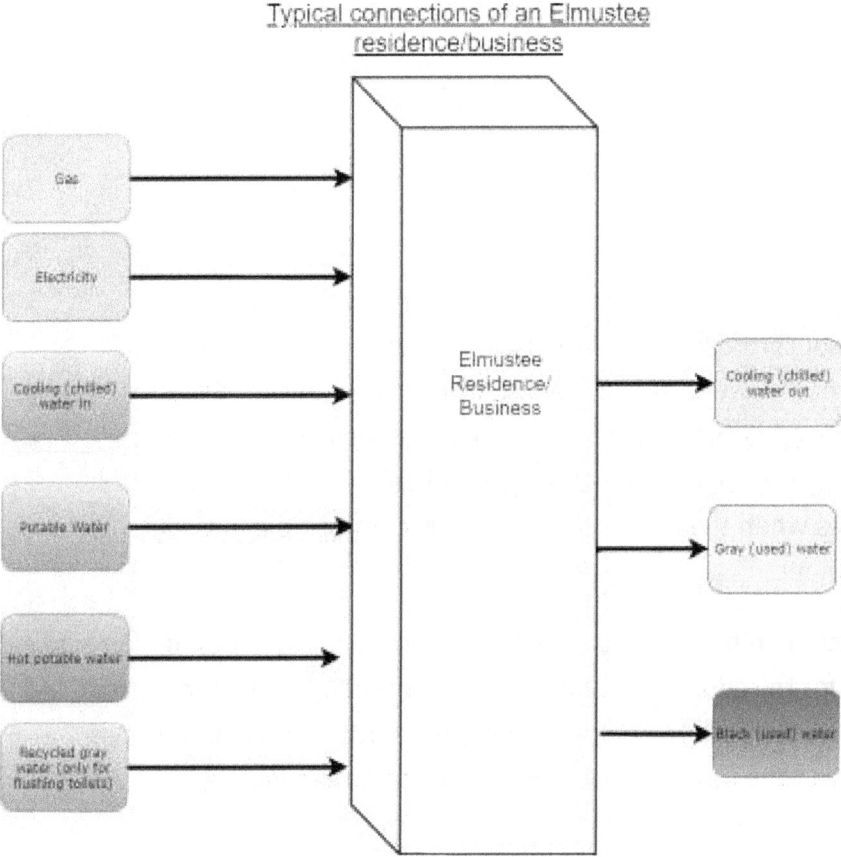

Figure 16 Elmustee residences and businesses utility connections

8 Elmustee's passive cooling systems

Synopsis

There is a very narrow band of ambient temperature human beings feel comfortable in. Most people report comfort in an environment that has an ambient temperature range of 65° F to 80° F. Temperatures outside this range are considered either too cold or too hot--people's motivation and efficiency to do work decreases when they don't feel thermally comfortable in their environment.

Construction in hot climates must incorporate passive cooling techniques to reduce the load on the air conditioning system. Elmustee uses passive cooling techniques, both at the community level and at the individual structure level, to better manage its energy resources.

Located in hot climate, at the community level, Elmustee wants to provide a naturally cooler environment to its residents. This goal is accomplished by employing various passive cooling technologies. The emphasis is on stopping solar insolation and tempering warm air coming from outside. Furthermore, air inside Elmustee is also cooled down for added general comfort.

Description

- State of the Art

The environment we operate in is composed of three elements: earth, water, and air. Of these three the air when heated or cooled gains and loses heat the fastest. If we block heat from reaching the air we are breathing in, we can create a naturally pleasant environment for ourselves. In an open space air is heated up by earth's top surface (which in turn gets heated up by the sunlight), and through contact with air coming from the surrounding hot areas. The passive cooling techniques in Elmustee work through isolation of Elmustee's ambient air from the aforementioned two influences.

- Technology

Elmustee uses passive cooling techniques at both the community and the individual structure levels. At the community level, the emphasis is on reducing the solar input to the land parcel by blocking sun during certain hours of the day during summer. Hot air moving in from neighboring areas into Elmustee, is intercepted by a green belt around the community. Inside Elmustee, evaporative cooling technique is used to bring the ambient temperature further down.

- Application

Reducing Solar Insolation component of heat

Elmustee uses operable curtains and photovoltaic panels to block off the sun during most day hours in the summer. East to West running streets in Elmustee have fifty-five feet high poles, twenty feet apart. Photovoltaic panels installed on these poles shade the community. All community buildings and commercial properties are located under this huge photovoltaic array.

Elmustee also uses operable curtains to block sun. Many East to West running streets in Elmustee have curtain-carrying fifty-five feet high poles, twenty feet apart. Curtains, 11 ft. high from the ground and up to the full height of the pole are drawn in and out along these poles to shade the community. Whereas many curtains are made of plain fabric, some curtains are made of thin film photovoltaic membrane to produce electricity. These curtains are drawn out between 10 am and 5 pm during the summer. The PV panels and the screens act as the roof of our community. Employing these techniques, the sunlight does not reach the ground, the ground does not heat up and in turn does not give its heat to the ambient air.

Blocking convection flows into Elmustee

Water and vegetation around Elmustee is our community's first defense against intruding convection currents from neighboring hot regions. First, Elmustee is isolated from its surroundings through a moat. The moat around Elmustee is filled with seawater. Seven row deep mangroves around the moat produce a cooling effect by tempering the warm air coming in the community from outside. The mangroves are watered using seawater; because of the slope provided around the moat the runoff ends up back in the moat. The vegetation at the moat acts as a wall for the community--this wall helps block hot air coming into Elmustee.

Evaporative Cooling on a large scale

Elmustee has several tall fences covered with thatched coconut leaves. Small pumps spray seawater on these fences. These thatched-leave fences soaked in seawater provide further evaporative cooling to the community. After every few days thatched-leave sheets are removed from the fence and dried in sun; the accumulated sea salt is recovered from this membrane.

Cooling trenches

Besides passive cooling as mentioned above, passive and active cooling is provided through concentric shallow circular ditches of water around the community, with water bodies closest to the community being machine-cooled. Through these cooling trenches the community is cocooned in a virtual cool-bubble. The cooling trenches are covered with wire mesh to prevent bug infestation. The refrigerated water is produced with the help of animal power.

First Cost

Costs of making the necessary components of the Elmustee Passive Cooling System are provided in a separate spreadsheet.

The contractor winning the contract for the operation of the Elmustee Passive Cooling System for the first three years will build the system using a loan from Elmustee.

Operating Costs

Maintenance and operating costs of the Elmustee Passive Cooling operations are being presented in a separate spreadsheet mentioned above.

Elmustee awards annual contract for the maintenance of its passive cooling system. The contractor is responsible for maintaining adequate level of seawater in the moat, irrigation of the mangrove forest, maintenance of the thatched-leave fences, and maintenance of the operable curtain system. The cost is covered from the annual membership dues.

Benefits

Using passive cooling techniques, Elmustee substantially brings down the air conditioning loads of the individual buildings in the community. Since all buildings--industrial, commercial, residential, community, and institutional--are cooled through district cooling, metered cooling-water provided to individual structures is a precious commodity. Elmustee's passive cooling strategy helps individual users conserve energy and save money for themselves.

A further benefit of the extensive use of evaporative cooling is realized in the generation of usable water. With air being saturated with water, traditional refrigeration and Warka towers produce copious amounts of freshwater. The water thus generated through condensation is added to the Elmustee freshwater reserves.

Conclusion & Specific Comments

Researchers are studying the impact of prolong shading on microorganisms' populations in day-time-shaded areas of Elmustee. Research results will be shared with the Elmustee members and the wider world.

Elmustee Passive Cooling

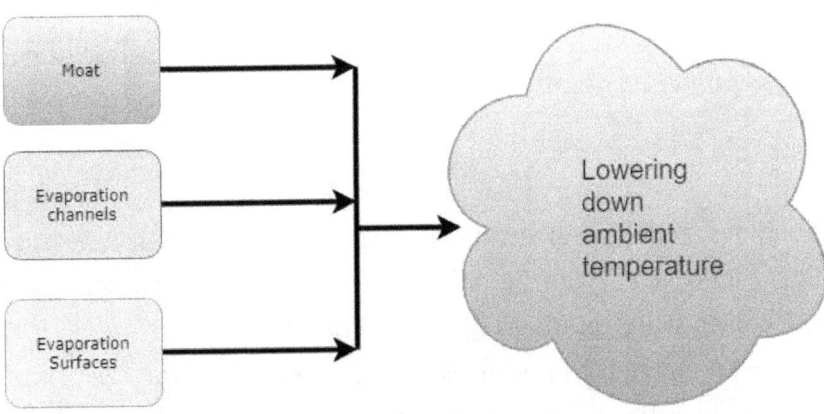

Figure 17 Elmustee passive cooling techniques

9 Elmustee's district cooling system

Synopsis

Elmustee buildings, though well insulated, still need active cooling. Air conditioning systems employed by the Elmustee buildings use chilled water pumped to them. The chilled water provided to each building is metered. Buildings are charged for their cooling expenses based on the amount of chilled water they use in a given month. Chilled water is generated using chillers that in turn use animal power--draft animals are used to operate refrigerant compressors.

Description

- State of the Art

In the climate zone Elmustee is located in, cooling needs of residences are generally met with wall-mounted air conditioners or split units. In large office buildings, cooling is often provided by central plants housing chillers and air handling units. Irrespective of the particular cooling system being employed in these buildings, electricity is used to power the refrigeration cycle. This reliance on electricity, a precious and often unreliable commodity in Pakistan, jeopardizes the smooth operation of these cooling systems. Moreover since electricity is mostly produced by thermal power plants in Pakistan, these cooling systems in fact are a big burden on the economy as oil, needed to run the thermal power plants, is mostly imported from other countries.

In places where wall-mounted air conditioners, or floor-mounted condensing units are used with several units present in a small place, the units fight each other for the rejection of heat resulting in decreased efficiency of the cooling systems.

District cooling is the process of generating chilled water away from the points of use. Using district cooling not only the heat rejection efficiency of the equipment is improved, the size of the cooling plant is also reduced based on the aggregate maximum cooling demand of several structures.

- Technology

The total cooling peak demand of Elmustee is calculated to be 2000 tons of refrigeration. Chilled water holding the required cooling capacity is generated at the duplex central plant (each part of the plant being of 1000 ton cooling capacity) using animal power. Draft animals work to operate, i) the refrigerant compressors (the backbone of the refrigeration cycle), ii) the cooling tower fans, and iii) the condenser water and chilled water pumps. The total demand of this cooling system is around 4,300 HP that is met by employing around 6,000 draft animals. Elmustee's district cooling system uses thermal energy storage tanks to store excess cooling produced by the animals. Twelve-inch diameter chilled water supply and return pipes meander through the community, supplying chilled water to each building and bringing warm water back to the central plant.

- Application

Elmustee buildings being well insulated have manageable cooling loads. Residences and other buildings use fan coil units and air-handlers. Chilled water--pushed by chilled water pumps--used in these forced-air units is supplied by the central district cooling plant.

At the district cooling plant all required mechanical power is generated through working draft animals. The central compressors are run by draft animals moving in a large circle. Pumps and fans are also operated using the same technique.

First Cost

Costs of making the necessary components of the Elmustee district cooling system (central plant with its chillers, pumps, etc.) are provided in a separate spreadsheet.

The contractor winning the contract for the operation of the Elmustee district cooling system for the first three years will build the system using a loan from Elmustee.

Operating Costs

Elmustee awards annual contract to a successful bidder for the maintenance and operation of its district cooling system. The contractor is responsible for doing the necessary repairs and keeping district cooling plant and the chilled water distribution system in working condition. The contractor also does the accounting work related to the cooling system, including billing the end-users for their chilled water usage.

Benefits

Using the district cooling strategy, the points of use of chilled water are physically separated from the generation point, thus increasing the efficiency of the district cooling plant. Using animal power for most of the refrigeration needs, Elmustee ensures keeping its electric demand low and its energy source local.

Conclusion & Specific Comments

At Elmustee Community active cooling (mechanical cooling) is only employed after making sure that the buildings have small cooling loads to begin with. District cooling is used to increase the efficiency of the refrigeration operation.

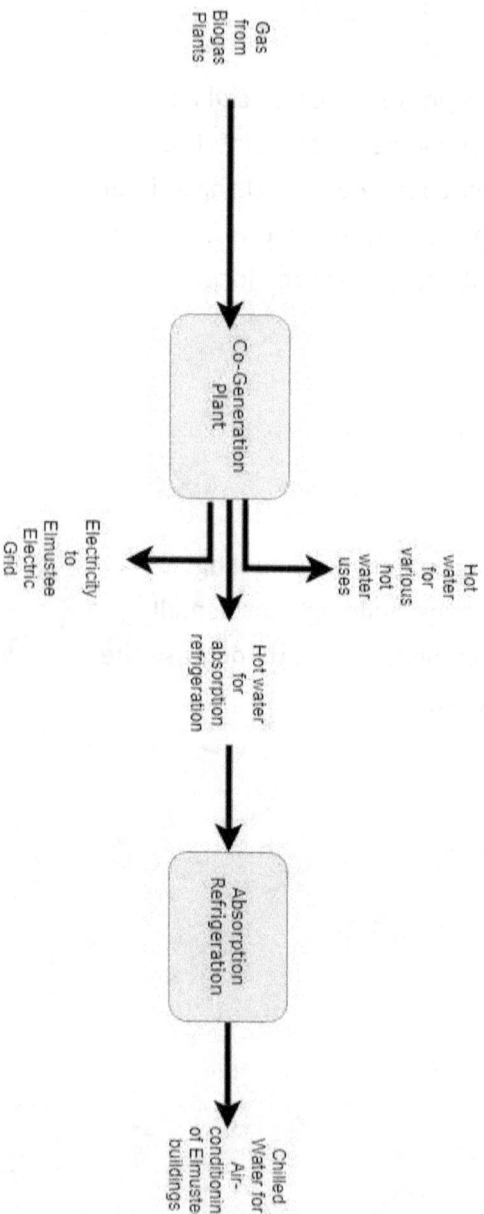

Figure 18 Elmustee co-generation plant—mechanical equipment (compressors, pumps, and fans) can be run by animal muscle power.

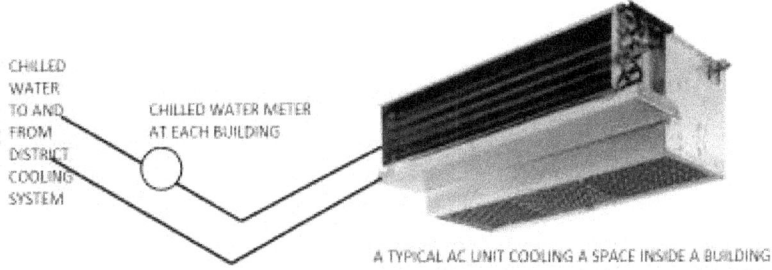

CHILLED
WATER
TO AND CHILLED WATER METER
FROM AT EACH BUILDING
DISTRICT
COOLING
SYSTEM

A TYPICAL AC UNIT COOLING A SPACE INSIDE A BUILDING

Figure 19 Building air conditioning system runs on chilled water supplied by the co-generation plant.

10 Rainwater harvesting

Synopsis
Modern urban settings have to deal with storm water runoff.
Elmustee makes maximum use of the water that rains on its land
parcel. All rainwater is harvested and channeled to above and
below ground water tanks. The stored water is later filtered and
sanitized before being made available for the use of the Elmustee
residents.

Description
● State of the Art
In modern built environments, rainwater falling on permeable
surfaces charges underground water reservoirs; rainwater falling
on hard surfaces is collected through a storm drainage system--
the runoff is then usually channeled to low elevation points. In
modern settings, rainwater is seldom collected for human usage.
At Elmustee storm water runoff over impermeable surfaces is
collected through the community's storm drainage system.
Rainwater falling over dirt and landscaping too is collected
through temporary-installed rainwater harvesting receptacles and
channels.

● Technology
At every Elmustee building with large roofs, rainwater is
channeled to a storm drain system, mostly directly, but
occasionally through water turbines--these small turbines add to
Elmustee's energy resources. Rainwater falling on small area
roofs and other hard surfaces is also collected and channeled to
various underground storage tanks scattered throughout

Elmustee. To collect rainwater showering on permeable surfaces (dirt and vegetation), a temporary system of tarp receptacles and flexible hoses is used. Rain collectors are 6'X6' tarp pieces, reinforced with aluminum strips at the periphery and sloped inwards to channel water to the center--at the center of the rain collector a 4" plastic sleeve facilitates a connection with a flexible 4"-8"-8" tee. Each rain collector is supported on the ground with a 12" high aluminum frame. Flexible hoses from rainwater collectors connect to a system of 8" plastic pipes temporarily put on the ground and sloped towards the nearest permanent storm water channel. Permanent storm water collection system is a chain of connected cisterns with a large underground storage tank at the very end.

During the monsoon season Karachi receives an average of seven inches of rain every year. Heavy downpour in a short time frame is not uncommon--Elmustee administrators use such downpours to stock the community's water reserves.

Even with these aggressive water harvesting techniques, Elmustee Water Works only collects 1.5% of the total amount of water needed for the community. Elmustee continually uses recycling techniques to meet its water demand.

- Application

An independent vendor, a company running Elmustee Water Works, operates the rainwater harvesting system. During short rains, only the permanently installed storm water system is used for rainwater harvesting. The vendor, always vigilant on weather forecasts, installs the temporary rainwater collection system during monsoon rains with the highest yields.

First Cost

Costs of making the necessary components of the Elmustee Water Works (including the costs of the rain harvesting system) are provided in a separate spreadsheet.

The contractor winning the contract for the operation of the Elmustee Water Works for the first three years will build the system using a loan from Elmustee.

Operating Costs

Elmustee awards an annual contract to a successful bidder for the maintenance and operation of all systems, including the rainwater harvesting system, falling under Elmustee Water Works. The contractor is responsible for doing the necessary repairs and for keeping the storm drain and collection system in working condition. The contractor also does the accounting work related to the Water Works, including billing the end users for their water usage.

Benefits

With rainwater harvesting and continuous water recycling Elmustee will mostly be using only the water it gets in the form of rains. This self-sufficiency in meeting its water needs will contribute to Elmustee's strength as an independent community.

Conclusion & Specific Comments

Active rainwater harvesting is crucial in meeting Elmustee's goal as a self-sustaining community. Researchers are studying the water table at Elmustee and compiling data about its seasonal rise and fall.

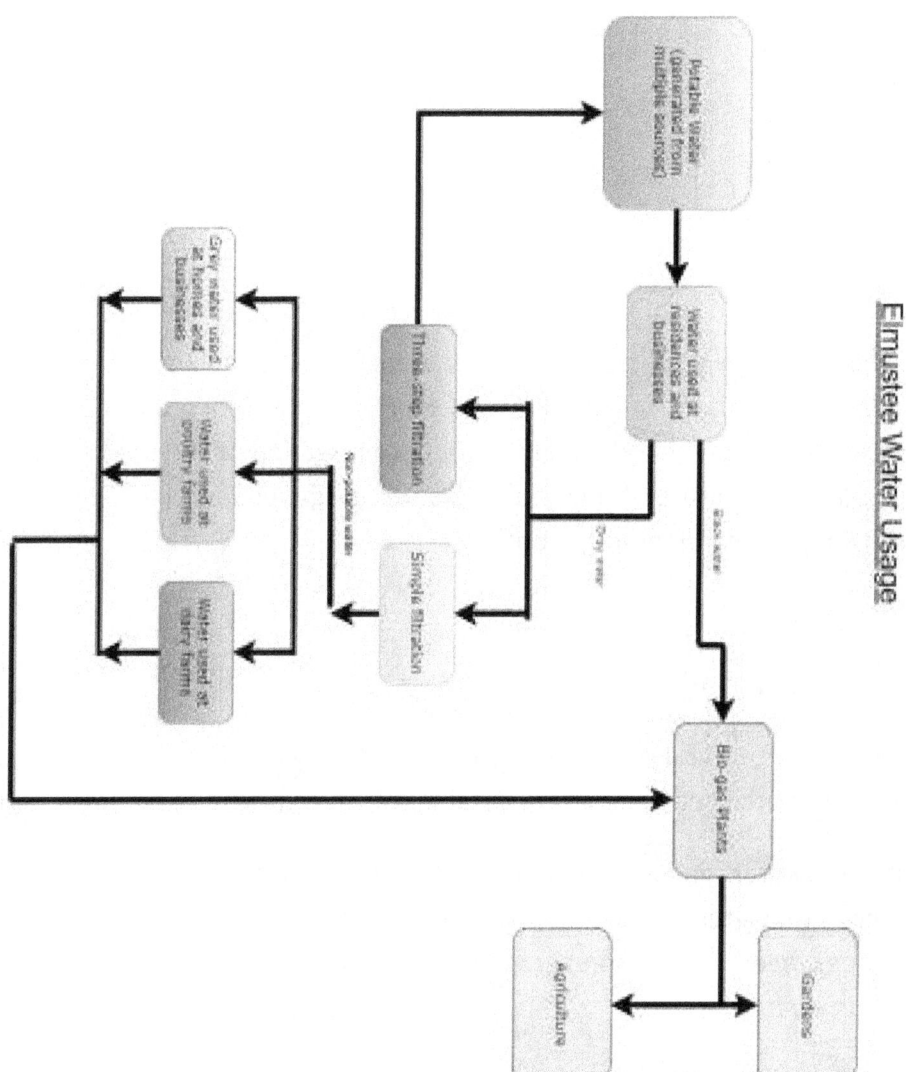

Figure 20 Elmustee water usage and recycling strategies

11 Seawater usage

Synopsis

At Elmustee seawater is used, a) to generate water suitable for human consumption, and b) to be mixed with grey water--see 'Water Recycling' section--and then used for irrigation of crops suitable for brackish water usage. Seawater usage reduces the freshwater demand for irrigation.

Description

- State of the Art

Reverse osmosis and reduced-pressure low-boiling point techniques are used to generate potable water from seawater.

Some seawater is mixed in equal part with grey water and used for brackish water irrigation--specific plants (certain types of potatoes, beets, dates, pomegranate, etc.) can be irrigated with brackish water.

Black water, collected from the water closets and urinals, is channeled to the biogas plant.

- Technology

At the RO (reverse osmosis) plants, animal power is used to generate potable water from seawater.

In parallel, potable water is also generated by boiling seawater at low pressure and then condensing the steam. At the low-boiling-point plant, pressure in the water containers is reduced by using vacuum pumps run by animal power. Vacuum is also generated in special gravity-tubes where after filling up the container to its maximum, the bottom of the container is dropped several feet generating vacuum on top of the water column. Seawater present in reduced pressure environment is heated using Fresnel lenses focusing sunrays on the water surface. Water vapors thus produced are condensed to obtain freshwater.

- Application

Using special pumps and CPVC pipes, seawater is pumped to the RO and Low-boiling-point plants.

Draft animals are driven in a circle; through gears, this animal power is used to generate pressure in the supply water. In the RO plant, potable water is collected on the low pressure side of the permeable membrane. The potable water generated by the RO plant is tested before being supplied for human consumption.

Seawater is also pumped to an array of 12 inch diameter, 8 feet tall metal tubes. Water going into these tubes is in fact sitting on top of a piston. When the tubes are filled all the way up, water supply is stopped and the pistons are dropped 3 feet. [The pistons are attached to a metal plate that is propped up at a height of 3 feet through a single metal column supporting the plate in the middle. Animal power is used to slightly lift the plate and then remove the column from underneath. When released the metal plate comes down under its own weight.] With this 3 feet of near-vacuum generated in the top portion of the tubes, the boiling point of water is reduced and boiling the water gets easier. At this point Fresnel lenses are used to heat up the tubes using sunrays. Steam generated in the tubes is removed by vacuum pumps. Steam is made to pass through a coil which is cooled using seawater going towards the metal tubes. Condensed water is removed from the coil to add to the potable water supply of the community.

First Cost

Costs of making the necessary components of the Elmustee Water Works (including the costs of making seawater purification system) are provided in a separate spreadsheet.

The contractor winning the contract for the operation of the Elmustee Water Works for the first three years will build the system using a loan from Elmustee.

Operating Costs

Elmustee awards an annual contract to a successful bidder for the maintenance and operation of all systems, including the seawater purifications systems, falling under Elmustee Water Works. The contractor is responsible for doing the necessary repairs and for keeping the seawater purification system in working condition. The contractor also does the accounting work related to the Water Works, including billing the end users for their water usage.

Benefits

Elmustee uses its proximity to sea to generate potable water from the seawater. This operation ensures continuous water supply for the residents.

Conclusion & Specific Comments

A 10" water line brings seawater to the moat--the same system also supplies water to the RO and low-boiling-point plants generating potable water for the community. Besides at the moat, the seawater is also used at the thatched-leaves-fence evaporative coolers.

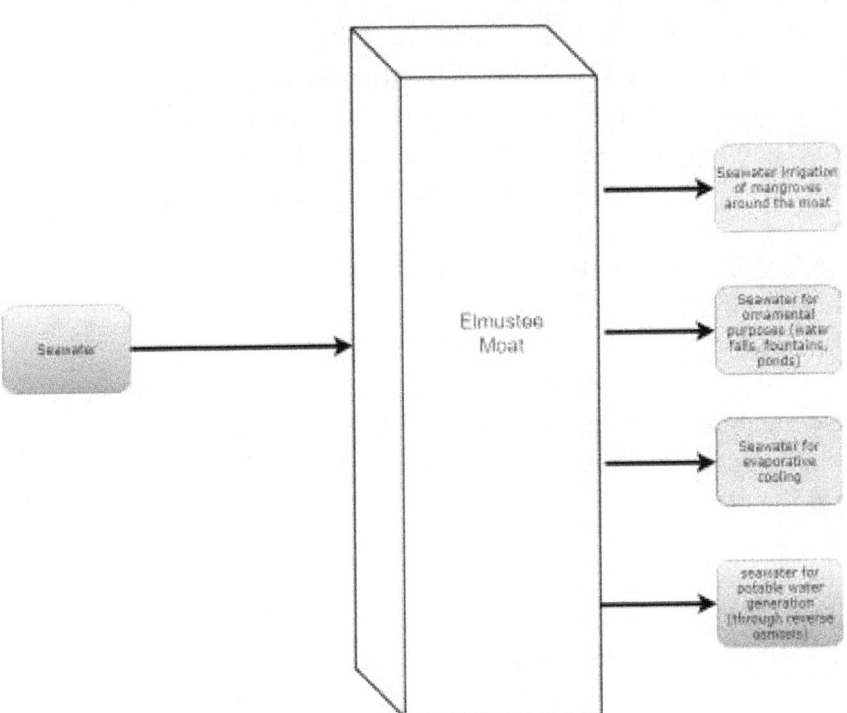

Figure 21 Seawater usage at Elmustee

12 Elmustee's water recycling system

Synopsis
Water is a shared commodity--through the natural water cycle, it keeps rotating between and among people and places. Elmustee is an active employer of water recycling technologies.

Description
• State of the Art
Water used in Elmustee buildings is segregated into gray and black water categories. Gray water, collected from sinks, lavatories, dishwashers, clothes washers, and showers is channeled to recycling stations where the gray water is, a) nominally filtered and supplied to the buildings to be used in the water closets and the urinals for the flushing needs, or b) thoroughly filtered, sanitized and sent back to the community buildings.
Black water, collected from the water closets and urinals, is piped to the biogas plant.
• Technology
At water recycling units, pumps are used to push water through sand and activated carbon filters, through RO (reverse osmosis) plants, and through UV filters. Draft animals are used for the rotation of the pumps.
Black water from Elmustee buildings is mixed with animal manure to feed the biogas plant.
• Application
Animal power is used to drive pumps that build water pressure; this pressurized water is used at the RO plant for the generation of potable water.

First Cost
Costs of making the necessary components of the Elmustee Water Works (including the costs of installing the water recycling system) are provided in a separate spreadsheet.

Operating Costs
Elmustee awards an annual contract to a successful bidder for the maintenance and operation of all systems, including the water recycling system, falling under Elmustee Water Works. The contractor is responsible for doing the necessary repairs and for keeping the recycling system in working condition. The contractor also does the accounting work related to the Water Works, including billing the end users for their water usage.

Benefits
Because of the limited amount of space available to harvest rainwater, and limited energy available to purify seawater, Elmustee actively employs water recycling techniques to ensure continuous supply of water to the community members.

Conclusion & Specific Comments
Water generated through rainwater harvesting and water recycling still does not meet the water needs of the community. As an added resource, Elmustee generates freshwater from seawater using reverse osmosis and low pressure boiling techniques. See 'Seawater usage' for further details.

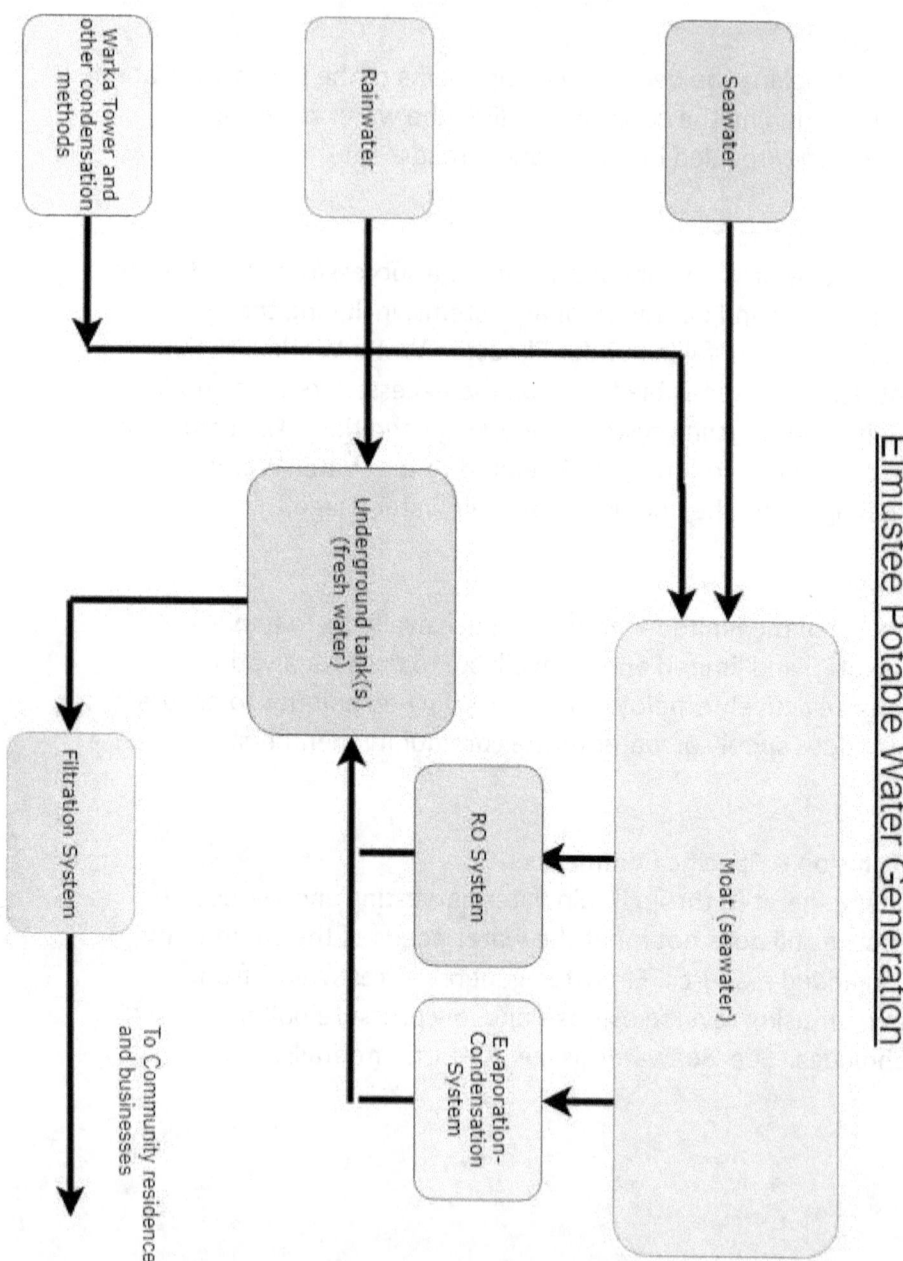

Figure 8 Water usage and recycling at Elmustee

13 Elmustee power generation system

Synopsis

At Elmustee, electricity is only used where, i) it is the most efficient energy source to use, and/or, ii) for a particular application, it is the only energy source that can be used. Our community members use electricity to power their electronic devices, small appliances and for lighting. At Elmustee, the air conditioning load, normally the largest consumer of electricity elsewhere, is taken care of through a district cooling system.

At Elmustee electricity is generated by

a) Photovoltaic (PV-solar) panels,

b) Windmills,

c) Animal power, and

d) Methane-powered electric generators.

Description

- State of the Art

Ever since electricity has been made available to the masses, it has become the most preferable source of energy to power equipment and appliances--even when it is not the most efficient source of energy to be used. Today, the world over, electricity is mostly generated by burning fossil fuels, through hydropower, and from nuclear power plants. Dams, build to make hydropower, contribute to environmental degradation. The world has limited supply of fossil fuels and the burning operation hurts our environment in multiple ways. Nuclear energy produces waste that is very expensive to deal with. Cognizant of these issues with conventional methods of energy production, the primary source of electricity generation at Elmustee is its large photovoltaic (PV) array. The PV panels generate electricity needed to meet the peak electric demand. Electricity-generating windmills are used when wind conditions are favorable for the use of those windmills. Considering the capricious nature of sun and wind, Elmustee considers it beneficial to use animal power, in an emergency, to generate electricity. The logic of using animal power for the generation of electricity is pretty straightforward: plants are already highly efficient in converting solar energy into food, and animals do a good job in digesting the food from the plants and converting the food to their muscle energy. Elmustee uses this muscle energy to generate electricity. Lastly, Elmustee keeps a large supply of methane as a backup power source. The methane can be used in electric generators suitable for its use.

- Technology

PV Array

Each 2'X4' solar panel generates 100 W. Ten thousand panels, covering 80,000 sq. ft., are needed to generate 1 MW. At Elmustee 160,000 sq. ft. area (a little less than 4 acres) covered by photovoltaic panels is used to generate 2 MW of electricity to meet the peak solar demand of the community. The campground and several small structures are present under the PV array.

Wind Mills

Elmustee uses four windmills, each of a maximum of 2 MW capacity.

Animal Power

As a backup, Elmustee maintains ten animal-powered electric generators.

Each animal-powered electric generator has a generation capacity of 5 kilowatt and uses ten (10) animals. These animals, tied to a structure, walk in a circle to move a large gear. A gear box is used to make the electric generator rotate at the recommended speed.

The ten (10) animal-powered electric generators are capable to meet the minimum critical demand of 50 kW for the community emergency needs.

On days when animal-powered electric generators are not needed, animal power is used to pump water to high elevation tanks. Water from the elevated tanks is used to generate a hydropower plant, in an emergency.

Methane-driven electric generators

Standard gas generators are used to generate electricity using Elmustee's stored methane gas reserves.

- Application

The contractor awarded the electricity generation project runs the various electrical generation and distribution systems.

The contractor is also responsible for billing the end users, and for proper maintenance of the equipment.

First Cost

Costs of making the necessary components of the Elmustee Power Generation System are provided in a separate spreadsheet (costs include the cost of installing the 2 MW PV Array, cost of installing the windmills, cost of installing animal-powered electric generators, cost of hydropower generators, construction cost of elevated water tanks, cost of methane powered electric generators, cost of switchgear, power distribution system, smart meters, etc.)

The contractor winning the contract for the operation of the Elmustee Power Generation System for the first three years will build the system using a loan from Elmustee.

Operating Costs

Elmustee awards an annual contract to a successful bidder for the maintenance and operation of the Elmustee Power Generation System.

The contractor is responsible for running the operation, maintaining various electricity generation systems, and the distribution network, taking care of any needed repairs, and taking care of the accounting (billing, collection, salaries) work.

Benefits

Elmustee Power Generation System is environmental friendly and carries minimum carbon footprint.

Conclusion & Specific Comments

Using various electricity generation techniques Elmustee takes care of the electric needs of the community. Surplus electricity can be sold to neighboring communities or the national grid.

Elmustee Electricity

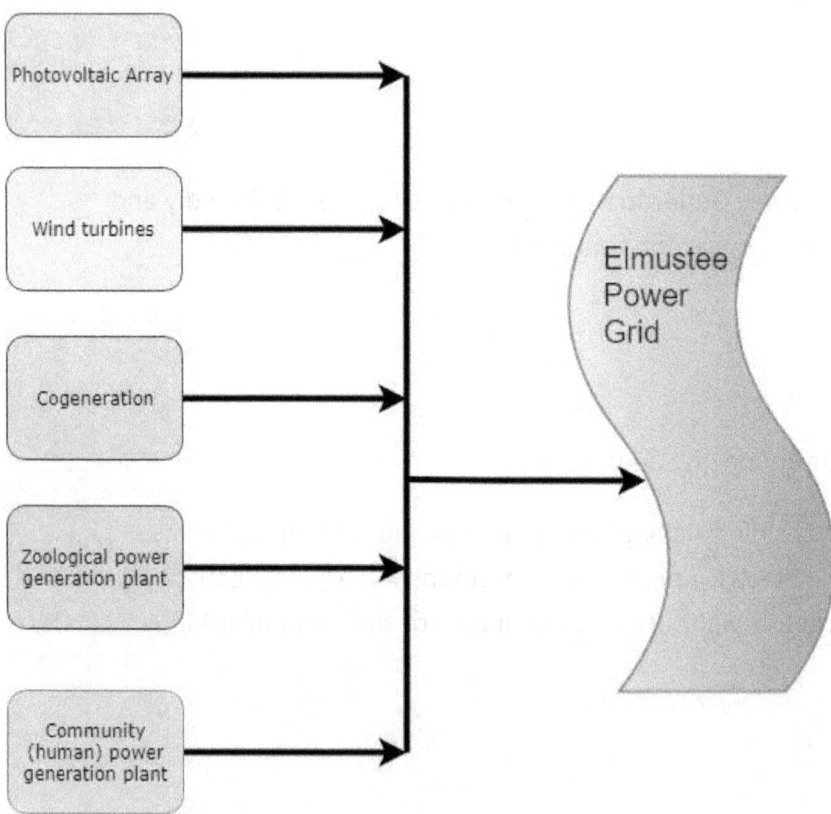

Figure 9 Elmustee power generation

14 Methane generation

Synopsis

At Elmustee three large biogas plants generate methane. Heating needs at Elmustee residences (kitchen and partial hot water generation) are met with methane. Methane-powered electric generators are used as backup power.

Description

- State of the Art

Methane is a big contributor to the greenhouse effect. Elmustee extracts maximum amount of methane from the generated waste and uses methane to meet the various needs of the community.

- Technology

Three separate biogas plants receive waste from the community. Waste water can be diverted to any of the three plants while shutting off a plant that may require maintenance.

- Application

All waste generated by Elmustee residents--people and animals-- ends up in one of the three biogas plants. Gas generated by the biogas plants is piped to, a) residences and businesses, to meet their cooking needs, and to b) a compressed gas facility where biogas is processed and filled in tanks to be used remotely.

First Cost

Costs of making the necessary components of the Elmustee biogas system (construction of three biogas plants, cost of installing a bio-methane filling plant) are provided in a separate spreadsheet.

The contractor winning the contract for the operation of the agricultural farms and the associated systems (including the biogas system) for the first three years will build the systems using a loan from Elmustee.

Operating Costs

Elmustee awards an annual contract to a successful bidder for the maintenance and operation of the Elmustee agricultural farms and associated systems (including the biogas plant and bio-methane system)

The contractor is responsible for running the operation and taking care of any needed repairs.

Benefits

The biogas plants at Elmustee help the community in meeting its goal of maximum resource utilization. The biogas plants generate methane and provide fertilizer-rich water vital for irrigating crops.

Conclusion & Specific Comments

Elmustee is constantly working on the automatic operation of the biogas plants.

.

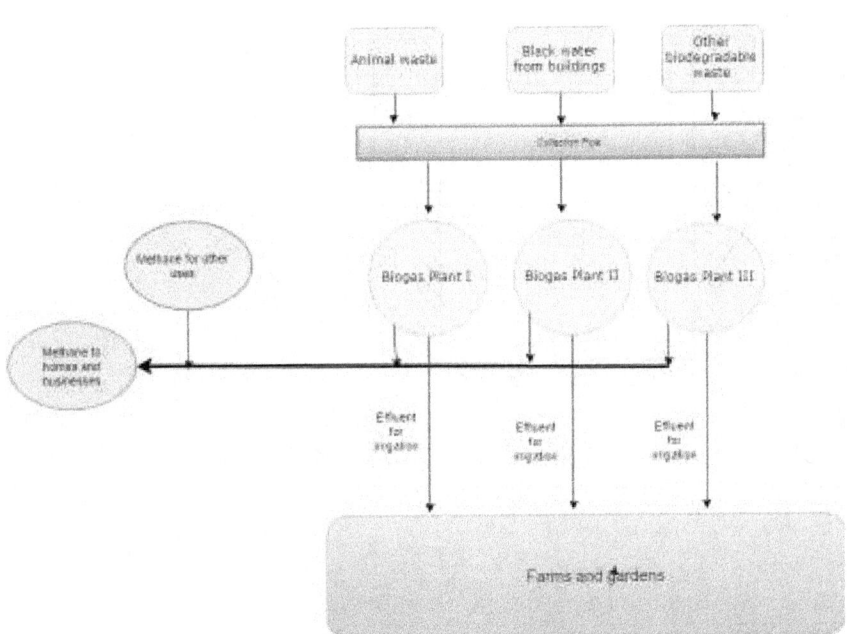

Figure 10 Methane generation at Elmustee

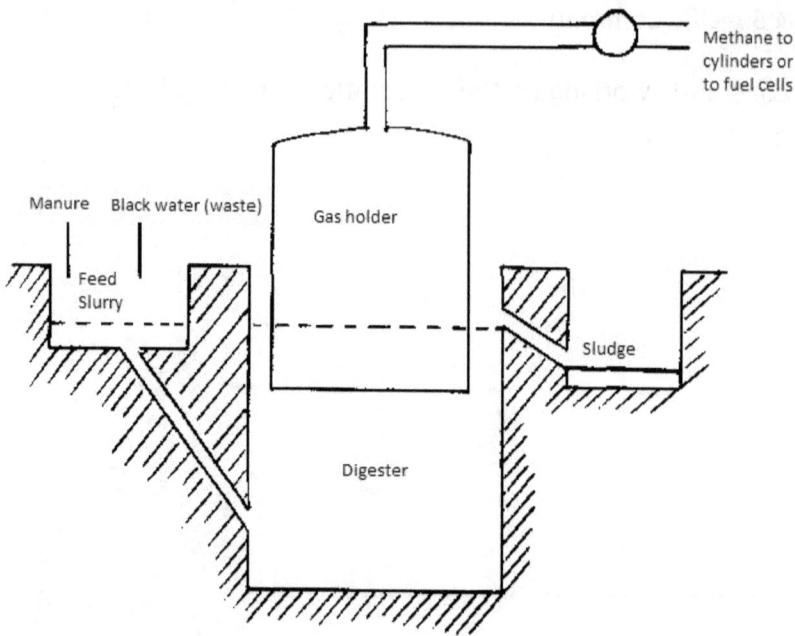

Figure 11 Biogas plant at Elmustee

15 Elmustee's hot water generation system

Synopsis

At Elmustee, hot water is primarily produced by the co-generation plant and by using solar energy. Methane hot water heaters are used as backup to the primary two sources of hot-water generation.

Description

- State of the Art

At residences and businesses hot water is used in the kitchen, and to meet washing needs. Outside Elmustee, it is common to use natural gas to generate hot water needed for human consumption. At Elmustee hot water is either produced in the co-generation plant or solar thermal panels and large Fresnel lenses are used to heat up water.

- Technology

Hot water generation plant, employing solar thermal panels and Fresnel lenses, is used to generate hot water to meet the community's needs. For hot water generated by the Co-generation system, please see 'Elmustee's district cooling system.'

- Application

Four large hot water generation plants, each generating around 100 gallons of water per minute at 120° F are used, at the Elmustee Hot Water Generation Plant.

First Cost

Costs of making the necessary components of the Elmustee Hot Water System are provided in a separate spreadsheet (costs include cost of making the large kitchen-greenhouse, cost of making twenty (20) Fresnel lens mercury heating systems, etc.)

The contractor winning the contract for the operation of the Elmustee Water Works (including the operation of the heating system) for the first three years will build the systems (including the hot water heating system and Elmustee Commercial Kitchen facility hot water system) using a loan from Elmustee.

Operating Costs

Elmustee awards an annual contract to a successful bidder for the maintenance and operation of all systems, including the hot water generation system, falling under Elmustee Water Works. The contractor is responsible for doing the necessary repairs and for keeping the hot water system in working condition. The contractor also does the accounting work related to the Water Works, including billing the end users for their water usage.

Benefits

Keeping hot-water generation activity away from the point of use, the community saves on the air conditioning requirements-- escaped heat does not contribute to the cooling load of the buildings.

Generating hot water at a central location reduces the demand on biogas being supplied to individual businesses and residences, for kitchen needs.

Conclusion & Specific Comments

Like all other energy system being used at Elmustee, the hot water generation system too is environment-friendly. Furthermore, Elmustee is constantly working on automatic operation of the hot water generation plants.

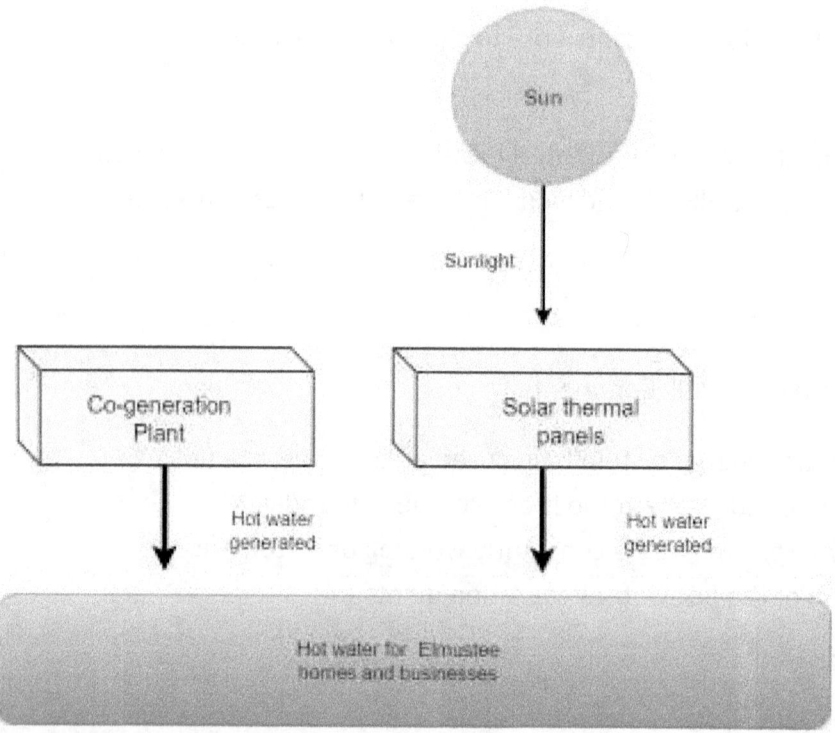

Figure 12 Hot water generation at Elmustee

16 Elmustee's commercial kitchen

Synopsis

The automated commercial kitchen facility at Elmustee uses solar energy to meet most of its heating needs.

Description

- State of the Art

Cooking, a chemical process, uses a lot of heat. Customarily, natural gas is used to meet a kitchen's heating requirements. At Elmustee, a community being built on the principles of complete self-reliance, mostly solar heating is used to prepare food in the community's commercial kitchen facility.

- Technology

A greenhouse, several Fresnel lenses, and a mercury-driven natural convection system are used in Elmustee's commercial kitchen facility.

- Application

Elmustee's commercial kitchen has a large south-facing greenhouse with glass top and south-facing glass; the bottom is made of insulated wood structure. All cooking operation takes place in a concentric glass box used in the bigger greenhouse. This arrangement, known as the kitchen-greenhouse, is accessed from the north through insulated glass doors. Using two concentric glass boxes, most of the solar insolation falling on the greenhouse is trapped. The internal glass box is connected to an exhaust fan; the fan is only run when the interior of the glass box fills up with vapors or food fumes.

Concurrent to the kitchen-greenhouse, Fresnel lenses heat up metal tanks containing mercury. As mercury is heated up in the tank it rises up and reaches coil-shaped heat-exchangers serving as 'burners.' Mercury, on losing heat to the food container placed on the burner, cools down and flows down to the metal tanks through another tube. Through heat collected by the Fresnel lenses, mercury is delivered to the 'burners' at temperatures around 300° Celsius. The heat trapped in the greenhouse aided by the heat provided by the burners helps cook the food.

The commercial kitchen has a methane-heating backup system for 'rainy days.'

The commercial kitchen uses filtered seawater for cooking food— the demand on freshwater is reduced, and the water already has salt in it; high heat cooking gets rid of any bacteria present in the filtered seawater.

Filtered hot seawater is used to clean pots, pans, and dishes.

First Cost

Costs of making the necessary components of the Elmustee Commercial Kitchen facility are provided in a separate spreadsheet (these costs include cost of making the large kitchen-greenhouse, cost of making twenty (20) Fresnel lens mercury heating systems, etc.)

The contractor winning the contract for the operation of the commercial kitchen heating system for the first three years will build the heating system of the Elmustee Commercial Kitchen facility using a loan from Elmustee.

Operating Costs

The commercial kitchen solar heating system is maintained by the same contractor that maintains Elmustee's hot water heating system. The annual contract includes operation and maintenance of the kitchen solar heating system.

Benefits

Instead of relying on fossil fuels, Elmustee's commercial kitchen relies on sun, the ultimate source of energy.

Conclusion & Specific Comments

The normal operation of Elmustee's commercial kitchen does not involve methane-based heating. Elmustee's precious methane supply--obtained from the biogas plant--is used in the kitchen facility only when the solar energy is not available.

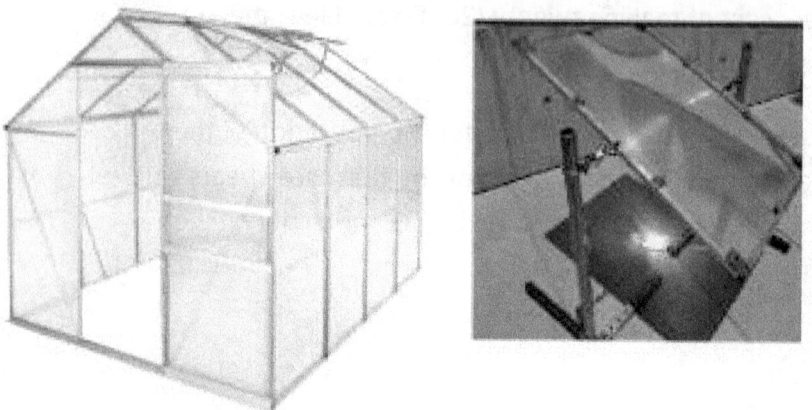

Figure 13 Elmustee commercial kitchen uses greenhouse and Fresnel lenses

17 Food vending machines at Elmustee

Synopsis

At Elmustee basic staple diet is made available at a low cost through vending machines that dispense dal, plain rice, and roti.

Description

- State of the Art

Vending machines provide necessary automation to food dispensing operation. But modern vending machines are seldom used to provide hot food, and especially hot desi food. Latest technology developed at Elmustee bridges the gap between desi food availability and vending machine convenience.

- Technology

At Elmustee specially designed vending machines dispense dal, plain rice, and roti. The dal and rice vending machines are gravity-fed. The roti vending machine uses lift gate mechanism--after every dispensation the window moves a slot.

- Application

Hot food vending machines are located near the commercial kitchen. The vending machines are automatically filled with dal, plain rice, and roti.

First Cost

Cost of making specialized vending machines capable of dispensing desi food is provided in a separate spread sheet.

Operating Costs

An outside vendor runs the commercial kitchen and the food vending machine operations.

Benefits

A mostly automated commercial kitchen and food vending machine operation provides healthy food at low cost to the Elmustee members.

Conclusion & Specific Comments

A mostly automated commercial kitchen coupled with automatic food vending machines is essential in providing low-cost staple food to Elmustee members. Such an arrangement fits well with Elmustee's social welfare policy of providing affordable food, shelter, education, and healthcare to all community members.

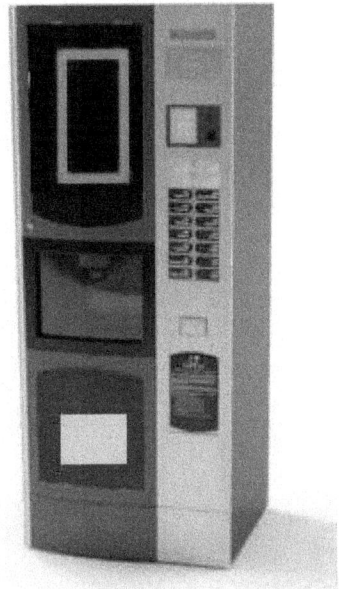

Figure 14 A dal vending machine at Elmustee

18 Elmustee health care facility

Synopsis

The Elmustee healthcare facility includes a 20-bed hospital along with medical offices, an OPD, a lab, and a pharmacy.

Description

- State of the Art

There is a growing trend towards using intelligent testing and diagnostic systems in the medical field. Employing computers at the very basic level reduces pressure on the medical staff.

- Technology

At Elmustee, latest testing and diagnostic medical systems are used to reduce involvement of full-time staff at the Elmustee healthcare facility.

- Application

Doctors and other staff members employed at the Elmustee Healthcare Center are members of the community.

As a part of Elmustee continuing education program, Elmustee Healthcare Center provides general first aid, CPR (cardiopulmonary resuscitation), AED (automated external defibrillator), and midwifery training to Elmustee members.

First Cost

Costs of building the Elmustee Healthcare Facility are provided in a separate spreadsheet.

Operating Costs

The Elmustee Healthcare Facility is run by a separate entity. The costs of running the healthcare facility are covered by offering health insurance to community members.

The monthly premium for health insurance depends on the frequency with which a member visits the healthcare facility.

Elmustee members are encouraged to keep an active lifestyle to reduce their risks of common and chronic illnesses.

Conclusion & Specific Comments

Besides serving the community members, the Elmustee Healthcare Facility also provides medical services to non-ambulatory non-members.

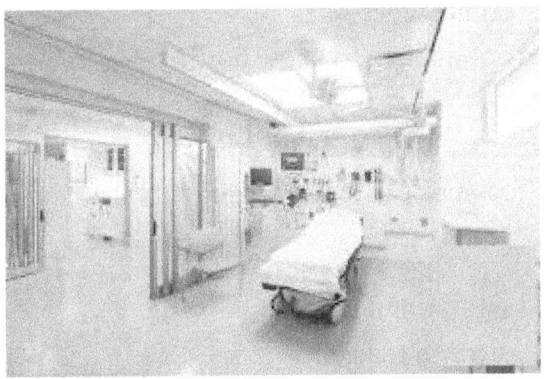

Figure 15 Elmustee health care facility

19 Education at Elmustee

Synopsis

Elmustee residents are provided free education up to the level equivalent to 10 years of schooling in the developed world. Further education is provided at workplaces where students earn while learning new skills.

Description

- State of the Art

The advent of the printing press was a monumental milestone in the history of mankind. Printing press paved the way for universal education. Only a few centuries after mechanized printing became a norm, civilized societies enacted minimum education requirements--all citizens must be able to read and write and have basic math skills. A complex world needs citizens capable of understanding complex concepts. Elmustee itself is a continuation of that thought--it is not enough to know how to read and write, and be able to add, subtract, multiply, and divide numbers; the social and environmental issues we face, require more education than what most societies call for, for their citizens to have.

After the popularity of the video cameras and lately of the Internet, the need to have a fully engaged teacher in the classroom is almost gone. Elmustee capitalizes on technology to provide universal education at a minimum cost borne by the community.

- Technology

At the Elmustee Public School pre-recorded videos are used in the classrooms. Once students reach the education level comparable to 10th grade in the developed world, and they are at least 16, further education is given at workplaces where students earn while they learn.

Testing centers organized by people associated with Elmustee certify students on knowledge and skills learned in specific fields.

- Application

Elmustee Public School mostly employs chaperons. These attendants keep discipline in the classroom and operate the necessary video equipment. The number of chaperons per 20 students decreases with increasing grade level.

Lectures are delivered in hologram. Students see the teacher right in front of them, writing and explaining concepts. After each lecture the students are tested on the material that was taught. Multiple hologram videos, recorded by different teachers, are used to explain the same concept in different ways.

Education beyond a level comparable to 10th grade in the developed world is given at workplaces where students work while they learn on their own, using education videos provided by Elmustee.

Elmustee does not have any two year or four year degree colleges. At Elmustee, a 'degree' in a specialized field is a specific combination of many certifications making up to that 'degree.' For example, a student interested in obtaining a four-year degree in 'Accounting' will first go through general education equivalent to 10 years of schooling in the developed world. After completing the general education, and being 16 or older, the student will start working for an employer doing business in the field of accounting. While working, the student will complete several short courses (in finance, audit, regulation, business, etc.) as a requirement for completing a 'four-year degree' in accounting-- proficiency in each course will be tested by a testing agency that will also issue a certificate to the student. While working, the students will be paid a minimum wage, with performance bonuses based on their grades in the specialized courses they take.

First Cost

Costs of constructing a school with several hologram lecture halls are provided in a separate spreadsheet.

Operating Costs

Elmustee mostly uses education material freely available in the public domain.

Testing and certification costs are borne by students/their employers.

Benefits

In today's fast changing world, people need to quickly acquire new skills. Elmustee's education system makes it easy for people to re-define themselves, at minimum cost to the community. By coupling higher education with work, students are provided a monetary incentive to get specialized education in their field of interest.

Conclusion & Specific Comments

Elmustee education system is continuously evolving to provide well-rounded education to the community members at the lowest cost.

Figure 16 Hologram teachers at Elmustee classrooms

20 Employment at Elmustee

Synopsis
Elmustee Community strives to be a place where residents are able to meet their basic needs at a low cost. Money is still the facilitator between needs and goods and services. Elmustee provides opportunities for people to earn a basic income to meet their food and shelter needs.
Elmustee community uses labor laws that in scope go beyond the laws used by the host country.

Description
• State of the Art
To the most part Elmustee administration is not involved in providing work opportunities to all its members. Community Administration's emphasis is on creating a healthy social structure where members meet each other's needs.
As a social service to the community, Elmustee provides basic work opportunities to its destitute members.
One way for any Elmustee member to make some money is to use their muscle power to generate electricity. The outdoor exercise machines let members charge their membership cards in exchange for the muscle power they use at these machines--the electricity generated with the help of the exercise machines charges batteries used as a backup to operate the food vending machines. Other exercise machine are used to pump water to an elevated location--later, the force of water, as it drops down, is used to generate electricity.

First Cost
Costs of making the necessary components of the Elmustee Community Power Generation System include the cost of

installing the exercise machines and the associated electricity generation or water pumping system are provided in a separate spreadsheet.

The contractor winning the contract for the operation of the Elmustee Community Power Generation System for the first three years will build the system using a loan from Elmustee. [See 'Elmustee Power Generation System']

Operating Costs
The annual contract awarded for the maintenance and operation of the Elmustee Power Generation System includes maintenance and operation of the exercise machines. [See 'Elmustee Power Generation System']

Elmustee Labor Laws
Employers working in Elmustee abide by the Elmustee labor laws. These laws require employers hiring labor to not only pay a minimum wage and to ensure the safety of the labor but to also provide for the personnel's physical well-being. As an example employers in Elmustee are required to arrange for shade for labor working in the sun for over 30 minutes. The labor laws describe the techniques and the strategies employers can use to provide shade for personnel working outdoors.

Figure 17 Exercise machines generating electricity

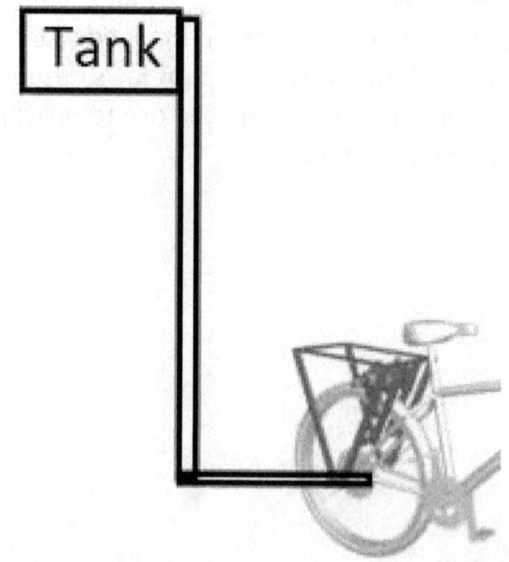

Figure 18 Human powered pump

21 The Elmustee community gardens

Synopsis

Community Gardens are located in a four story building within Elmustee. Resident and non-resident members can grow vegetables, flowers, and other plants on garden plots they rent.

Description

• State of the Art

Elmustee members are provided opportunities to grow their own plants/food. Garden plots are rented out in a four story structure.

• Application

The community gardens building is fashioned in the form of a single-helix (sloped) structure. Sunlit areas of the structure can be used for growing most crops; 'darker' areas of the structure are used to grow mushrooms and other plants that can be grown in the shade.

Large mirrors reflect light inside the multi-story community garden.

First Cost

Cost of constructing a single-helix four-story community garden structure is included in the cost table.

Operating Costs

Manpower associated with operation and maintenance of the community garden structure is also included in the cost table.

Conclusion & Specific Comments

Elmustee engineers are working on a plan to provide mechanized movement of the garden plots, for better opportunities for the garden plots to share sunlight.

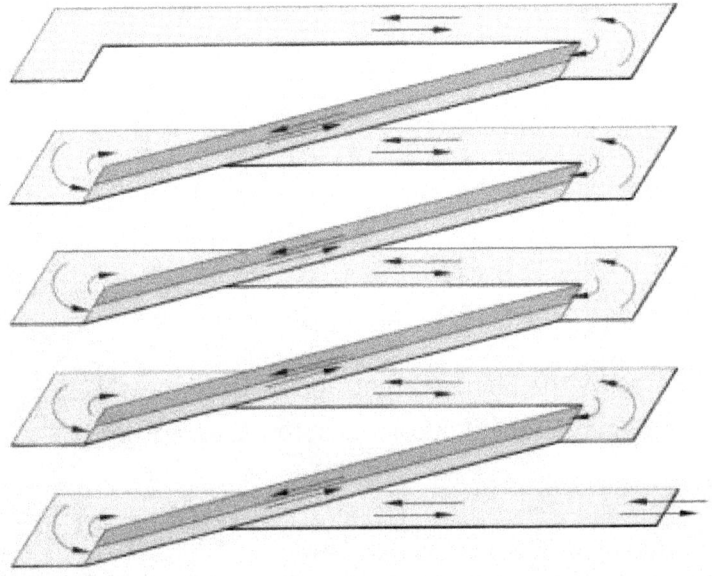

Figure 19 Elmustee community gardens

22 The Elmustee storage facility

Synopsis

A storage facility located by the Elmustee entrance is used mostly by non-resident members to store their bicycles, tents, and other belongings.

Description

- State of the Art

Elmustee does not allow outside vehicles to enter its premises. Walkways and bikeways provide healthy alternatives to transportation by cars. Non-resident members find it convenient to store their bicycles, skateboards, and other sporting gear at Elmustee.

- Application

The storage facility is a part of the border control building. Small and large storage bins are rented on daily, weekly, and monthly basis. Rental income adds to Elmustee's income stream.

First Cost

See costs under 'Membership, Identification, and Border Control.'

Operating Costs

See costs under 'Membership, Identification, and Border Control.'

Conclusion & Specific Comments

Since Elmustee members are charged by weight, any luggage they bring into Elmustee, non-resident members find it useful to store their bicycles and other stuff at the Elmustee Storage.

Figure 20 Self-storage facility near the Elmustee entrance

23 Law enforcement at Elmustee

Synopsis
Even with active surveillance of all public areas in the community, Elmustee needs a police force. Elmustee security force doubles as the community police.

Description
- State of the Art

Because of the restricted use of the community and the presence of video and audio recording devices and sniffing apparatus throughout the general areas in the community, a strong deterrent to crime exists. Still, any untoward incident is handled by the Elmustee security force.
- Application

Elmustee does not have a police station on its premises. Anyone detained by the security personnel is handed over to the area police. Elmustee maintains an active relationship with the law enforcement agencies of the area.

First Cost
See costs under 'Elmustee General Security'.

Operating Costs
See costs under 'Elmustee General Security'.

Conclusion & Specific Comments
Elmustee actively facilitates membership of law enforcement personnel of the area. See the section on 'Membership, Identification, and Border Control.'

Law enforcement at Elmustee

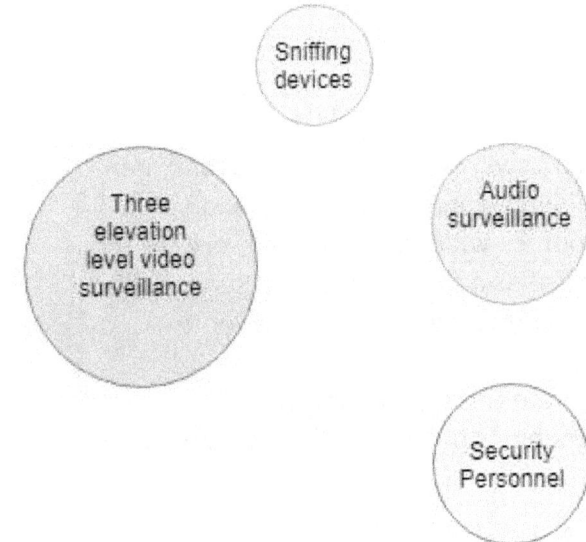

Figure 21 Law enforcement structure at Elmustee

24 Elmustee communication systems

Synopsis
Whereas Elmustee has its own wireless phone service between reception, community buildings, campground area, and residences, commercial country-wide mobile phone services too work at Elmustee. Elmustee has a satellite internet service as a backup, but mostly uses a commercial cable service for its internet use.
Elmustee radio and TV operate through the Internet.
Elmustee web site is community's portal for all services provided by Elmustee.

Description
- State of the Art
There are no wired phone services at Elmustee. Elmustee members use their own cell phones that are served by country-wide service providers. Elmustee residences and businesses use internet cable service provided by an outside service provider.
In the spirit of keeping an essential communication system completely free of outside interference, Elmustee maintains a basic wireless system between its four main areas: the reception, the community area, the campground, and the residential area. Similarly, Elmustee subscribes to a satellite internet service as a backup for its internet needs.

- Application
Elmustee is served by a cable service providing internet connection services to the community members and businesses. Satellite dishes at the reception, the library building, and the office at the residences provide the backup satellite internet option.

News at Elmustee is provided through Internet-based Elmustee radio and TV networks.

First Cost
Costs of making the necessary components of the Elmustee communication system are provided in a separate spreadsheet.

Operating Costs
Elmustee awards an annual contract to a successful bidder for the maintenance and operation of the communication system.
The contractor is responsible for keeping the system in good condition and for taking care of any needed repairs.

Conclusion & Specific Comments
With improving technology, the backup internet and wireless phone services are being constantly upgraded.

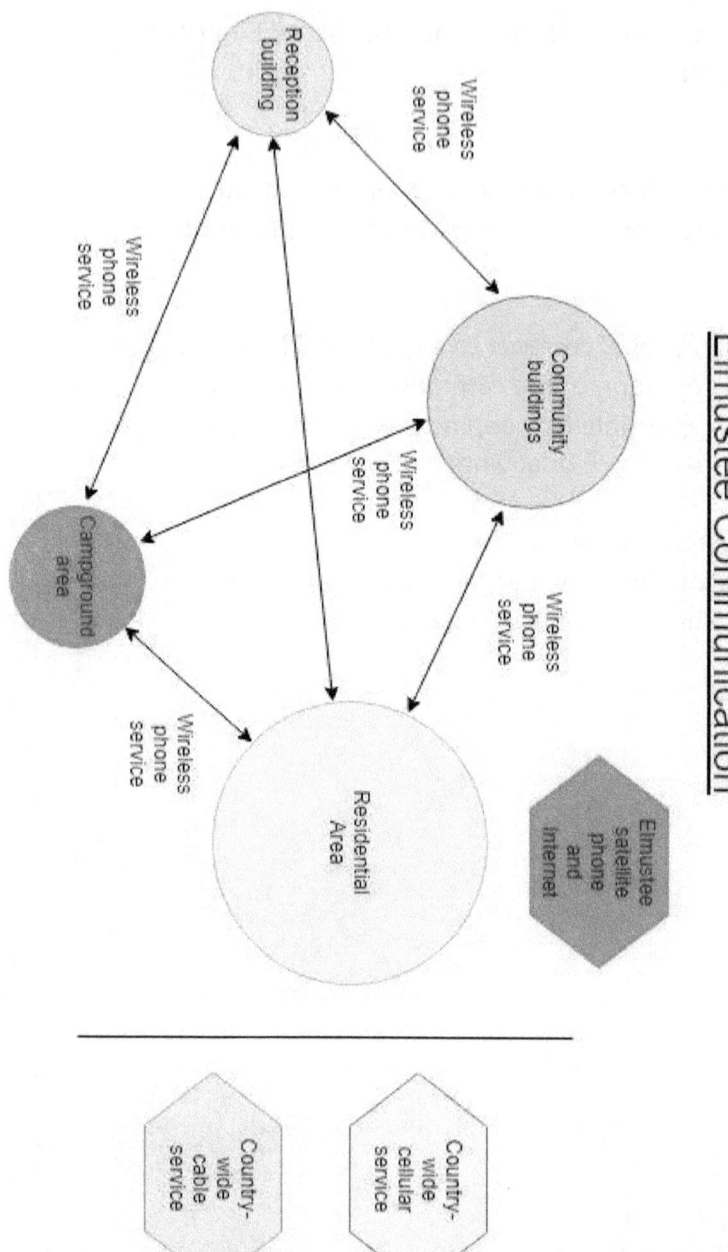

Figure 22 Communication systems at Elmustee

25 The campground and the basic living facility at Elmustee

Synopsis

The campground and the associated living facility provide low-cost, short-term living arrangements for non-resident Elmustee members.

Description

- State of the Art

Most of the Elmustee members are non-residents, making short-term trips to the community. Because of the political instability of the environment Elmustee is located in, non-residing members often find it necessary to spend a night or two at Elmustee. Whereas full service privately-owned hotels are present inside the community, Elmustee facilitates low-cost short-term stay of its members through a campground and an associated basic living facility.

- Technology

An office located outside the fenced campground and living facility area collects necessary rent and reprograms a member's ID card for them to enter the facility. Entrance-exit privilege to the campground and living facility area is revoked at the end of the period a member's ID card is enabled for.

- Application

At the campground concrete pads are rented out for 24-hour durations. Members can bring their own tents, or rent them at the office. Area round the pads is mulched.

The living facility structure has plastic bunk beds. Thatched-leaves mats, mattresses, and sheets are included in the rent charged on per bed basis.

Renters are expected to leave the rooms in the same good condition they were given in, at the time of check-in. Any damage done to the room or the furniture is charged back to the renter.

Toilet and shower facilities are provided at a structure next to the campground.

First Cost

The contractor winning the contract for the operation of the campground for the first three years will build the campground, the living facility, public toilets, office, storage, etc. using a loan from Elmustee. The work includes leveling, fencing around the campground with turnstile (enabled with card reader), building concrete pads, building living facility structure, building toilets and shower structure, building office and storage structure, construction of washing and laundry facility, etc.

Operating Costs

Elmustee awards an annual contract to a successful bidder for the maintenance and operation of the campground and the living facility. Money collected as rent is used to maintain the facilities.

Benefits

The campground and the living facility provide opportunities for Elmustee's non-resident members to stay at Elmustee for longer durations, at low cost.

Conclusion & Specific Comments

The campground and the living facility are favored by Elmustee members ready to rough it out. Members requiring more comfort stay at any of the various commercial hotels at Elmustee.

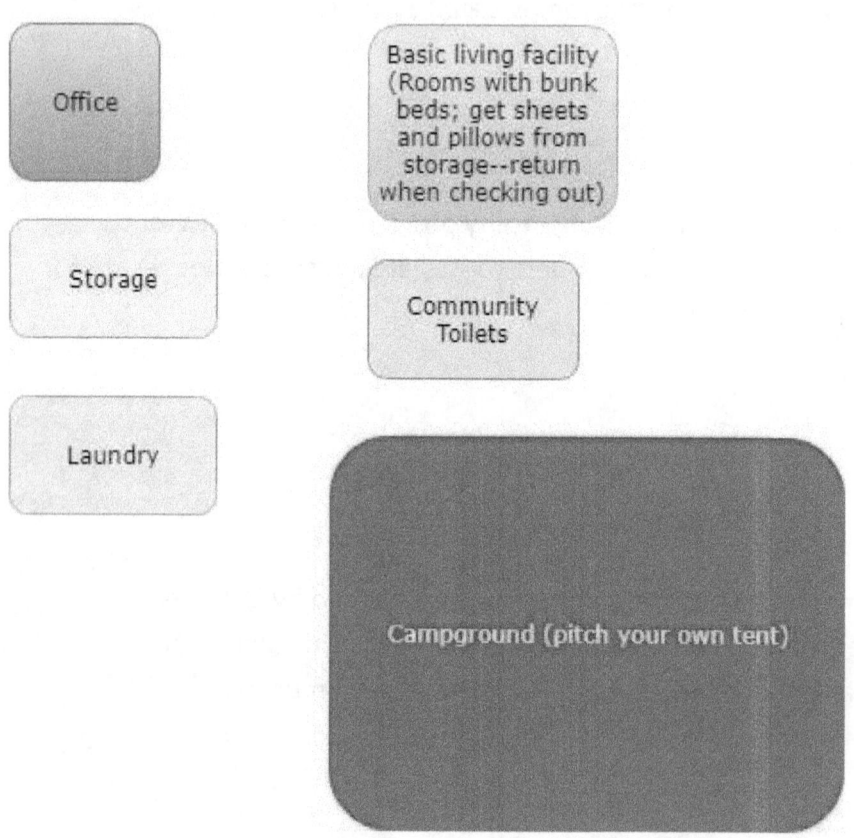

Figure 23 Campground and basic living facility components

26 Landscaping at Elmustee

Synopsis

Most of the Elmustee landscaping requiring freshwater is located under the shade of the PV panels. Some landscaping is irrigated using brackish water; a tiny portion is given greywater. The landscaping irrigation system is fully automated.

Description

- State of the Art

Trees and plants are essential in creating a beautiful living environment. The sun is vital for the growth of grasses and plants, but too much sunlight unnecessarily increases up the evaporation rate from the vegetation. At Elmustee, landscaping features are either present around the moat where they are watered by the seawater, or are present under the shade of the photovoltaic panels where their evaporation rate is greatly reduced.

- Technology

At Elmustee freshwater is considered a precious resource. Used freshwater is continually recycled. Consequently, landscaping that needs copious amount of freshwater has limited scope at Elmustee. Vegetation around the moat (mostly mangrove) uses seawater; and trees and plants sporadically used in the spaces between the entrance, the community area, and the residences are suitable for brackish water irrigation. Limited landscaping, present under the PV panels, uses greywater.

- Application

Seawater pumps draw water out of the moat to water plants around the moat.

Brackish water pumps draw water out of the brackish water tanks and water plants suitable for brackish water usage. This automatic irrigation system works in the middle of the night to reduce evaporation losses.

Recycle-water-pumps draw water out of the grey water tanks and water plants and grasses under the PV panels. This automatic irrigation system too works in the middle of the night to reduce evaporation losses.

Elmustee walkways and bikeways are mostly covered with vines.

First Cost

Costs of making the necessary components of the Elmustee landscaping are provided in a separate spreadsheet.

The contractor winning the contract for the maintenance of landscaping for the first three years will install the landscaping features using a loan from Elmustee.

Operating Costs

Elmustee awards an annual contract to a successful bidder for the maintenance and operation of the Elmustee landscaping.

Benefits

Evaporation from plants and grass is reduced by placing most of the landscaped areas under the shade of the PV panels. Mowed grass and plant trimmings are fed in the biogas plant.

Conclusion & Specific Comments

Like all other systems of the community, landscaping at Elmustee too is well-integrated with other Elmustee operations.

Landscaping at Elmustee

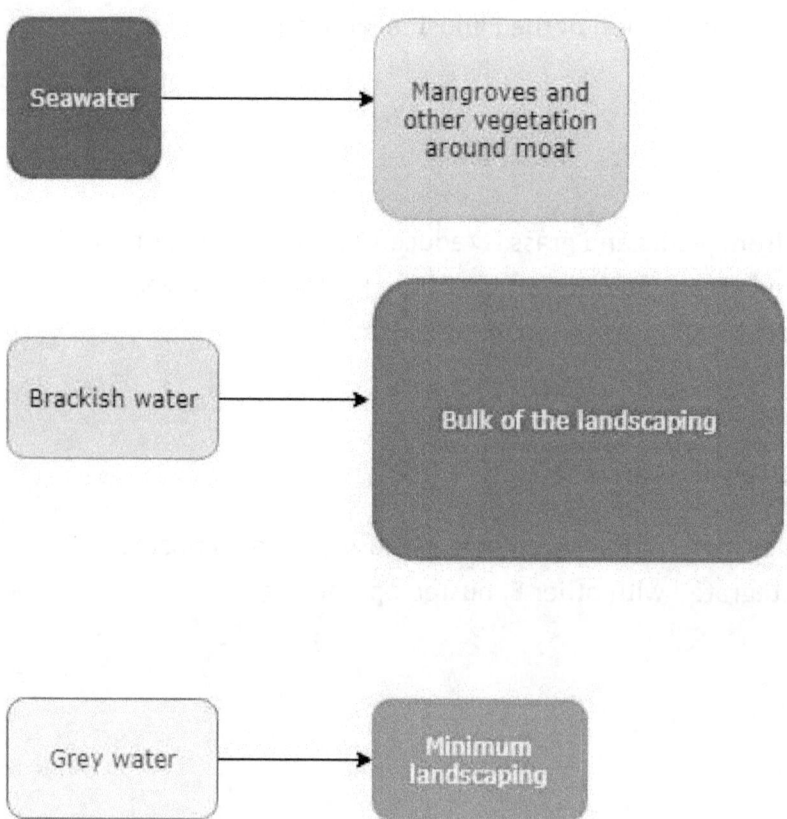

Figure 24 Landscaping water usage at Elmustee

27 Public toilets at Elmustee

Synopsis

Public toilets at Elmustee operate on a pay-per-use basis. Recycled water is used at the toilets; black water from these facilities is taken to the biogas plant.

Toilets are maintained by a contractor awarded yearly contract by the Elmustee administrators.

Description

- State of the Art

Public toilets and showers are present at the reception, at the community buildings, and at the campground. Entrance to the toilets is facilitated through turnstiles connected with card readers.

- Technology

Elmustee members use their ID cards at card readers to gain access to the toilet facility. At each scan a fixed fee is charged and the money is deducted from the ID card.

First Cost

First Cost related to the public toilets are included in the individual construction cost of each structure.

Operating Costs

Elmustee awards an annual contract to a successful bidder for the maintenance and operation of the public toilets.

The contractor is responsible for the proper maintenance of the public toilets.

Conclusion & Specific Comments

Men and Women public toilets and showers help Elmustee's non-residing members enjoy their visit to the community.

Public Toilets at Elmustee

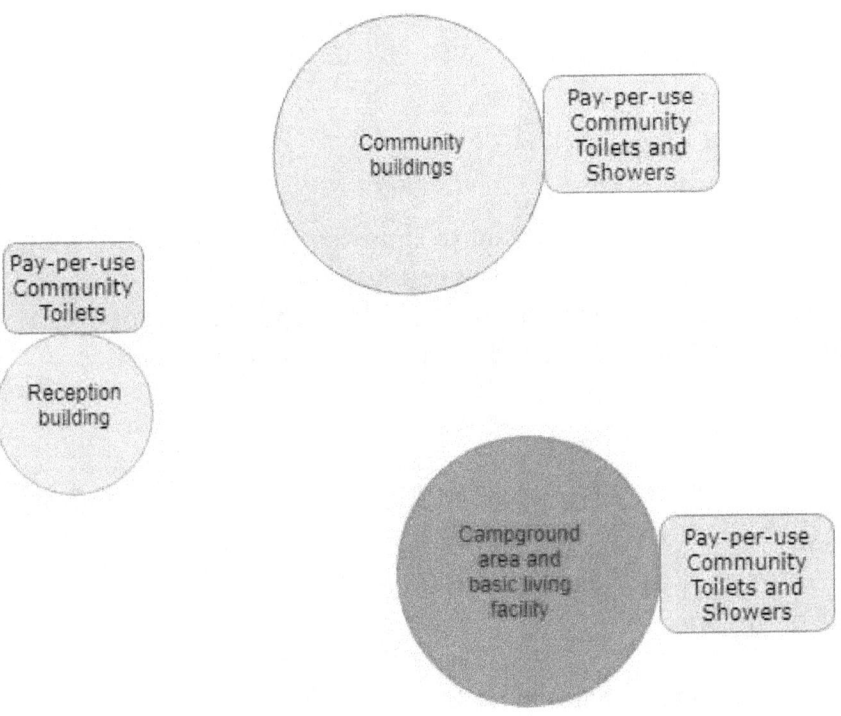

Figure 25 Public toilets and shower facilities at Elmustee

28 Food security and agriculture at Elmustee

Synopsis

Food security is extremely important to Elmustee. Elmustee grows grains, vegetables, and fruits at nearly automated farms. Food grown at the Elmustee farms meets the basic food needs of our residents, and the feed and fodder needs of our poultry and dairy animals.

Description

- State of the Art

Elmustee farms are totally automated. The agricultural farms are mechanically and often remotely tilled, sowed, irrigated, and harvested. The emphasis is on organic farming.

- Technology

After acquisition of land for Elmustee, farms were graded. Now automated farm machinery is used at each step of the farming operation: for tilling, for sowing, for irrigation, and for harvesting. Black water coming out of the biogas plant is used for irrigating the crops.

Robot-driven tractors are used to till the land. For smaller farms, cable tillers are used. [With cables running the length of the field, the electric motor-driven tiller moves on the cable and tills the land.]

In our community located by the sea, floating farms (on large rafts) using solar sills are used to grow food.

Climate-controlled greenhouses are used to grow trees and plants suited for various climates. [Greenhouses simulating tropical weather are used to grow various spices including clove, cardamom, cinnamon, etc.]

The micro-climate of humid environment near the fuel cells is used to grow plants that thrive in moist air.

• Application

For a population of 10,000, Elmustee needs to grow food and fodder on around 500 acres. Elmustee farms provide two crops annually, with a yield of around 7 tons per acre. Farms growing alfalfa provide 6 to 8 crops in a year. Besides being used internally in the community, harvest from the Elmustee vegetable and fruit gardens is exported to increase Elmustee's revenue stream.

Elmustee farms and gardens are irrigated by a) slurry (mixture of water and fertilizer) obtained from the biogas plant, and b) brackish water obtained by mixing seawater with treated greywater.

First Cost

Costs to initially prepare the almost automated Elmustee farms and gardens are provided in a separate spreadsheet.

The contractor winning the contract for the operation of the farms and gardens for the first three years will build the farms using a loan from Elmustee.

Operating Costs

Elmustee awards an annual contract to a successful bidder for the maintenance and operation of the Elmustee farms and gardens.

The contractor is responsible for running the operation, taking care of any needed repairs, and supplying (selling) the products to Elmustee and outside retailers.

Benefits

Though Elmustee does not grow everything that the community members need, Elmustee does grow grains, vegetables, and fruits to meet a bulk of the Elmustee food demand. This food self-reliance helps keep Elmustee retain its independence.

Conclusion & Specific Comments

Elmustee farming is well-integrated with the biogas plant and water systems.

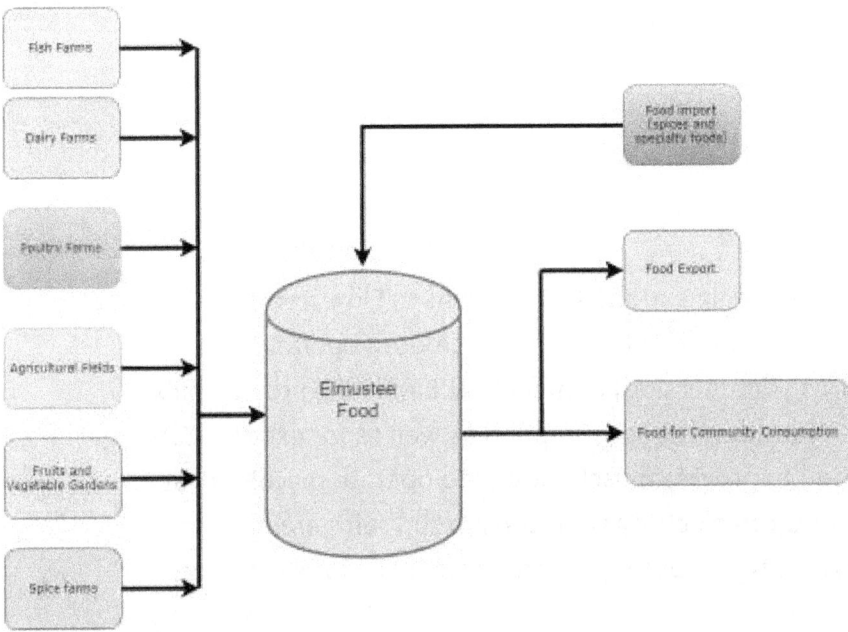

Figure 26 Elmustee Food Production

29 Dairy farms at Elmustee

Synopsis

Food security being extremely important to Elmustee, the community has special interest in the smooth operation of its dairy farms. Cattle, raised humanely at Elmustee, provide milk and meat for domestic consumption as well as for export. Cattle is also used to provide muscle-power to operate machinery at Elmustee: to run electric generators, to run refrigerant compressors, to run pumps at the Reverse-Osmosis plant, etc.

Description

- State of the Art

In the last thirty years there have been many developments in automating dairy farming. England and Australia have been in the forefront of renovations in the dairy farming industry. Elmustee builds on the successful experiments done in the West.

- Technology

In building the dairy farms, Elmustee uses the latest available technology. Elmustee dairy farms are to a great extent automated. Herds are let out and in the barns, using robots. Cows are milked using the well-tested robotic milking systems. Feed is also provided using an automatic system. The barns are regularly cleaned using an automatic washing system; the manure is pushed towards the biogas plant.

- Application

Elmustee keeps over 200 bulls and over 2000 cows to meet its muscle power and dairy needs. The bulls are employed at the power generation plants, the district cooling plant, and the RO plant.

First Cost

Costs of making the necessary components of the Elmustee Dairy Farms are provided in a separate spreadsheet.

The contractor winning the contract for the operation of the dairy farms for the first three years will build the farms using a loan from Elmustee.

Operating Costs

Elmustee awards an annual contract to a successful bidder for the maintenance and operation of the Elmustee Dairy Farms.

The contractor is responsible for running the operation, taking care of any needed repairs, and supplying (selling) the products to Elmustee and outside retailers.

Benefits

Elmustee uses its bulls to run the power generation plants. Waste from the farms is used in the biogas plant, to generate methane.

Conclusion & Specific Comments

Elmustee's automated dairy farms provide a cost-effective solution to meet the animal muscle power needs, and meat and milk requirements of our community.

Figure 27 Automated dairy farm at Elmustee

30 Poultry farms at Elmustee

Synopsis

Chicken, ducks, and turkeys are raised humanely at Elmustee to provide eggs and meat for domestic consumption as well as for export.

Description

- State of the Art

To a large extent Elmustee poultry farms are automated. Feed is provided through a mechanized system; the coops are inclined, making the laid eggs to roll down to a covered channel; and a dog-robot rounds up the chicken for them to go back to their coops at night time.

- Technology

Chicken coops are present in large cages where the wire-meshed walls are covered with vines, and the concrete floor is covered with hay. Chickens can come out of the coop, into the larger area any time--they are raised cage-free.

Similar arrangements exist for ducks and turkeys as well.

- Application

Poultry feeders are filled through pipes bringing the food from the top. Water continuously runs through a water channel located in the middle of the giant cage. Dog looking robots round up the chickens when needed. On certain days, all chickens are rounded up to go to an area attached to their cage. Once the cage is empty, blowers blow hay away to one end where the cage wall is automatically removed for the soiled hay to end up in a dumpster. Once the hay is removed water is sprayed on the floor and small robots are used to scrub the floor. Water is sprayed multiple times to ensure the floor is completely clean. Once the scrubbing robots are out, air jets dry up the floor. At this point nozzles located at the ceiling level spray hay to cover the floor. The chickens are let back in, once the cleaning work is complete.

A water-filled ditch around each cage acts as a deterrent to mice intrusion.

First Cost

Costs of making the necessary components of the Elmustee Poultry Farms are provided in a separate spreadsheet. The contractor winning the contract for the operation of the poultry farms for the first three years will build the farms using a loan from Elmustee.

Operating Costs

Elmustee awards an annual contract to a successful bidder for the maintenance and operation of the Poultry Farms. The contractor is responsible for running the operation, taking care of any needed repairs, and supplying (selling) the products to Elmustee and outside retailers.

Benefits

Elmustee poultry farms, run mostly automatically, administered by a vendor, provide organic eggs and meat for the consumption of the Elmustee members as well as for export.

Conclusion & Specific Comments

Elmustee is continually looking into making its poultry farms more automatic and more humane.

Figure 28 Free range automated poultry farm at Elmustee

31 Solid waste management at Elmustee

Synopsis

Elmustee uses the proven reduce-reuse-recycle strategy to deal with its solid waste. Waste reduction is encouraged through tiered waste collection charges--the less waste you generate the more you save on garbage disposal fee. Residents and businesses sort their garbage at source.

Description

- State of the Art

Elmustee residences and businesses sort their garbage at source. Garbage is sorted in the following categories: a) plastics, b) metals, c) glass, d) paper and cardboard, e) compost, and f) composites/general garbage. Residents and businesses pay for their sorted garbage to be hauled away. Minimum charges apply for picking up the recyclable or biodegradable garbage. Of the six categories, highest charges apply to 'general garbage'--high volume of general garbage attracting higher fee.

- Technology

Elmustee Solid Waste Management Company handles the solid waste generated at the community. Elmustee residents and business owners are educated through videos on how to sort the garbage at source. Six plastic bins, each clearly marked with the type of solid waste it is appropriate for, are provided to each household and business. The size of the bin determines the garbage collection fee the resident or business will pay.

Recyclable waste (metal, glass, and paper) is hauled to recycling facilities. Compost is used to make fertilizer. General garbage is buried in a landfill.

- Application

Garbage inspectors review the contents of bins before letting a golf-cart-truck haul it away. Penalty applies if a waste receptacle meant for one kind of garbage is found to be containing a different kind of waste.

First Cost

Costs of making the necessary components of the Elmustee solid waste management system are provided in a separate spreadsheet.

Operating Costs

Elmustee awards an annual contract to a successful bidder for the maintenance and operation of the Elmustee Solid Waste Company. The contractor is responsible for collecting the recycled waste and for disposing of general waste at the landfill site. The contractor also does the accounting work related to waste management, including billing the end users for their garbage collection.

Benefits

By educating Elmustee members on ways to reduce, reuse, sort, and recycle their solid waste, and by penalizing increased solid waste generation, Elmustee hopes to greatly reduce its solid waste problem.

Conclusion & Specific Comments

Elmustee is keeping an eye on its landfill site. Elmustee administrators may, at a later stage, decide to use an external landfill site far away from the community. Elmustee is also reviewing a plan to install a rail track to automatically move the bins to the recycling sites and the landfill, and to take empty bins back to the residences.

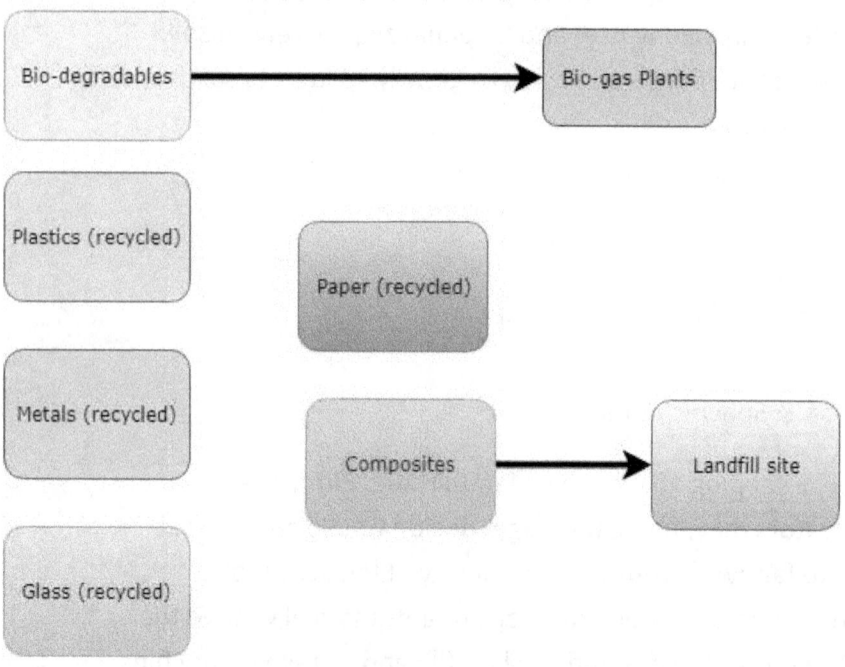

Figure 29 Elmustee waste stream

32 Elmustee Social welfare services

Synopsis

Elmustee believes in taking all its members along and paying special attention to those that are left behind for whatever reason.

Description

Elmustee strives to be a community where basic needs of all its members are met. At the same time Elmustee does not believe in handouts. To keep their self-respect intact, Elmustee has devised programs where an 'unemployed' member can 'work' for the community and earn money to pay for their food and shelter.

- Application

The center of the Elmustee social services is the community's power generation system. An 'unemployed' member of the community can work at the community power generation system and get compensated—their membership card is credited and they can use the card to pay for food and shelter at Elmustee.

Food vending machines at Elmustee provide basic food at a minimum price.

The Education Center at Elmustee helps people learn new skills.

Elmustee Library provides space for members to read, and to connect to the Internet.

Elmustee Health Center provides subsidized services to its members in need.

First Cost

Costs of various systems discussed above are presented under chapters discussing the individual components in greater detail.

Conclusion & Specific Comments

Elmustee believes a healthy society depends on all its members being useful and working. But Elmustee also recognizes how some people can be left behind. Elmustee is ready to help its members in need without creating a system where people are rewarded for not working.

Elmustee Social Services

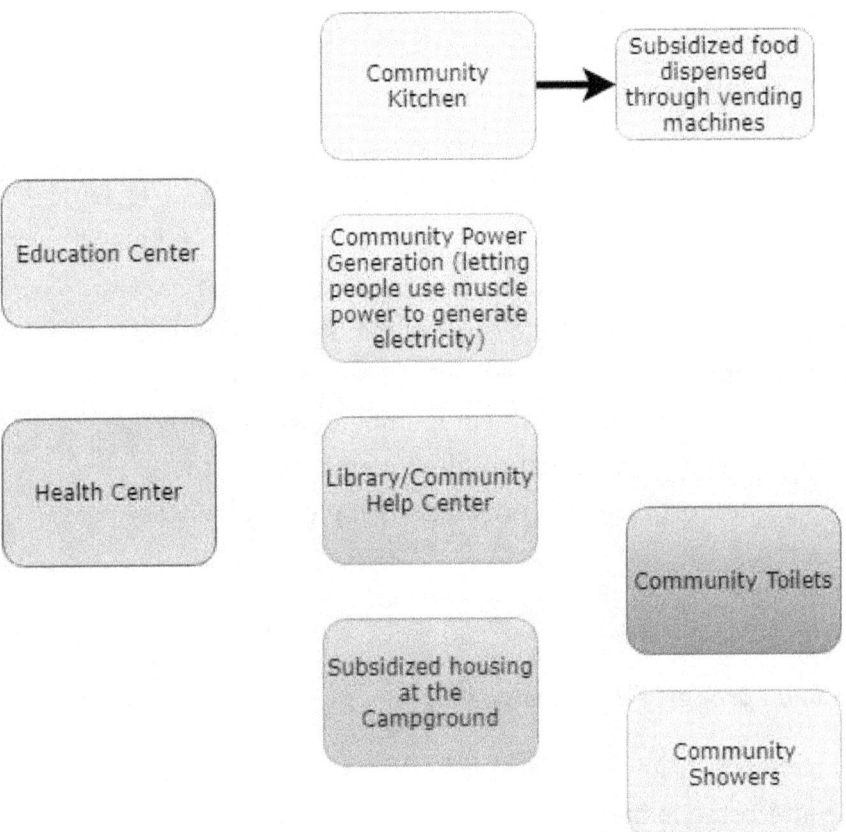

Figure 30 Social services at Elmustee

Frequently asked questions

Why is Elmustee being created?

Every country deserves a place that is accessible to all AND is well-run. Elmustee is that place. It is a gated community that is only open to its members: there are resident and non-resident members of this community. Resident members permanently live within the gated community; non-resident members visit the community for a short duration. Anyone can apply for Elmustee membership.

How can I buy property in Elmustee?

Property ownership rights in Elmustee are open to members only. If you are interested in buying a property in Elmustee you will first have to become a member of this community. When inducted in Elmustee, a member is issued a non-resident Elmustee membership card. When a member buys a property or establishes their residence in Elmustee, they are issued a resident membership card.

How do I become a member of Elmustee?

Anyone can become a member of Elmustee by

1. Filling out an online form (providing reference of two current Elmustee members) and paying the membership fee.

2. Passing a literacy test.

3. Going through Elmustee reading material and the orientation videos.

Why should I become a member of Elmustee?

Every Elmustee member has their own reasons for joining Elmustee. Some join to escape to a well-managed place, others come to enjoy the green acres of the community, and still others want to become part of a progressive movement.

Can I find work at Elmustee?

There are jobs available at Elmustee. Elmustee vendors are required to only hire people that are already members of Elmustee. The vendors continuously advertise part time jobs. These jobs are normally filled by new members who want to explore opportunities in Elmustee.

All Elmustee jobs pay in Elmustee currency. The basic pay—the minimum wage--helps a new member pay for their food and basic living facility at the campground.

What are the resident and non-resident areas in Elmustee?

There are no non-resident areas in Elmustee. We do have areas restricted for the use of residents only. Everybody, residents and non-residents, have access to all community areas.

Residents-only area in Elmustee provides privacy to Elmustee residents.

Why cannot I bring my motorized vehicle (car, motorcycle, etc.) in Elmustee?

Members can enter Elmustee only on foot, or on wheelchairs. Heavy items are brought in Elmustee using the railroad. The primary reason for this arrangement is our members' security. We can scrutinize people entering Elmustee better when they are without their vehicles. Moreover, Elmustee is a green community; members of Elmustee are very conscious of the quality of air they breathe. If outside vehicles are allowed to come in, Elmustee will lose control over the emission performance of the vehicles being driven in Elmustee. That is another reason why Elmustee does not allow non-Elmustee vehicles on its premises.

In Elmustee most people either walk or take the train to their destination. Motorized low-occupancy vehicles are available for emergency purposes.

Elmustee is very accessible. Ramps are provided at the buildings, and almost all places in Elmustee are accessible on a wheelchair.

Why is garbage such a big issue at Elmustee? Why is luggage brought from outside weighed and heavily charged at the Elmustee entrance?

Elmustee has limited capacity to recycle waste. Every commodity used by people, sooner or later enters the waste stream. And that is why Elmustee eyes things brought from outside with concern. Members pay for commodities brought from outside by the articles' weight because they are paying for the garbage Elmustee will have to deal with, in future.

Why is there a quota on males and females in Elmustee?

Elmustee Community wants to achieve a natural demographic balance: 50% women, 50% men. To make up for the past grievances against women we are ready to have a major share of them in Elmustee. Men are often asked to wait for admission in the community, when, at a given time, the men to women ratio inside the gated community gets to as high as 50-50. As an example, say there are currently 1000 female and 999 male members present inside Elmustee. A group of three men arrives. Only one man from this group can be allowed in at this time, to keep a 1:1 male to female ratio in the community. The other two men will be allowed in when either, i) two female members arrive, or ii) a group of men and women arrives and the group has at least two more women than the men, or iii) two men leave the community.

Why men are discouraged from walking in groups larger than 2?

Because some of the Elmustee female members have made this recommendation.

I am in the process of immigrating to Australia. Meanwhile, can I come and live in Elmustee?

If you have fulfilled Elmustee membership criteria and you become a member, you can come and live in Elmustee. Non-resident members can stay at privately run hotels or at low-cost campground and other living facilities.

At Elmustee you will find the same basic discipline, and civic life you are expecting to find in a well-run place. You may decide to settle here instead of moving to Australia.

What is Elmustee Daera?

The Elmustee Daera is a scheme for the world to have a sustainable use of its natural resources. People and communities fulfilling the Elmustee Daera criterion get the majority (more than 90%) of their food and energy needs met with resources available within 10 miles of their residence. Why 10 miles? Because a normal bicycle rider can easily make a trip of up to 10 miles to buy things she would use.

Step by step Elmustee

1. Idea is presented to investors. Investors create a company that will build Elmustee. Funds are secured for all expenses.

2. A security company is established. Personnel hired for the security go through the basic Elmustee training.

3. A piece of land is bought.

4. On acquiring the land, the security company moves to the parcel escorting personnel and equipment needed to do the initial setup. The initial setup will include the following:

   A. Trailers with bunkbeds.

   B. Photovoltaic panels, windmills, generators, gasoline and compressed gas cylinders.

   C. A trailer full of pipes, cables, and various tools.

   D. A nursery trailer with plants (big and small), and a large seed bank.

   E. A trailer with kitchen, pantry, and food supplies.

   F. A trailer with various farm animals and their feed.

   G. A trailer with a water filtration system.

5.   On reaching the acquired land the security personnel will make bunkers at the boundary and start the guard.

6.   Boundary of the community will be set up, and an entrance/exit point will be decided. From this moment on, no one will be able to enter the community from anywhere else but at the designated entrance.

7.   Waterwells are dug.

8.   Photovoltaic farms are set up.

9.   Zoological power plants are set up.

10. Utility lines are set up throughout the land parcel.

11. Agricultural fields, vegetable and fruit gardens are set up.

12. Simultaneous efforts are made to recruit members.

13. Construction work is carried out and model homes and offices are built to get reservations for future construction, for purchase by the members.

14. Vacation homes are built and are rented out, to start an income stream. [Renters have to meet the membership criteria as well.]

15. Residential towers and office buildings are built.

16. As members start buying properties, the investors are paid off.

17. Members start running the affairs of Elmustee.

Elmustee community will have around five thousand (5000) residents. Among these residents we will have all the skillsets needed for the community to be self-contained--from teachers to technicians, from architects to construction workers, the workforce will include everyone needed to run a community of this size.

Location: We have not bought a piece of land yet, but the plan is to have this community by the sea.

Right now we are recruiting members: people who will eventually be either the residents of the community, or its frequent visitors. Elmustee is a membership community; only members will be able to enter the premises. Unlike other gated communities, Elmustee will have a large number of non-resident members: people who would like to just visit and enjoy the amenities of a well-planned, well-run community.

Once the membership reaches 1000+, we will invite investors to invest in the project. The Elmustee Development Community, comprising construction professionals, will buy and develop the land--the members will buy the developed properties in the community. Ultimately the affairs of the community--on financial exit of the investors--will be run by the Elmustee members.

Phasing out the Elmustee construction work

Phase I

- Electricity infrastructure

- Security infrastructure

- Water supply and drainage infrastructure

- Construction of vacation and model homes

Phase II

- Extension of infrastructure

- Setting up a campground

- Construction of living spaces for 1000 people

- Construction of administrative offices, library, commercial office spaces

Phase III

- Finish up all construction work

- Living spaces for 5000 resident members

The roadmap

The initial one-thousand Elmustee members are invited based on the founders' personal contacts. Elmustee membership will later expand based on the community's criteria for membership—the criteria includes referral by two current members.

The Elmustee Development Company (a for-profit entity) will:

1. Buy the land

2. Build residences, commercial and community areas.

3. Build moat, boundary, and community security systems.

4. Install power generation units and energy distribution systems.

5. Build water reservoir and the water purification and distribution systems.

6. Build the drainage and water-recycling systems.

7. Carry out other construction work.

The Elmustee Development Company will recover its costs and make profit by selling residential, commercial, and industrial properties in the membership club. The ownership of the community will thus gradually move to the Elmustee Membership Community i.e., to its resident and non-resident members.

Once Elmustee is up and running, the Governing Body of the Elmustee Membership Community will award annual contracts to companies to run its following systems:

1. The Security system

2. The Power generation and distribution system

3. The Water distribution and drainage system

4. The General Maintenance system (including landscaping)

5. The Food farms

6. The waste management system

7. The transportation system

What Elmustee stands for

First and foremost, Elmustee is about security. It is a fortified membership club. All public areas in Elmustee have 24-hour video surveillance. This small community is striving to be called the world's best crime free society.

Elmustee is about a citizenship focused society. We extend our membership only to those who meet the community's criteria of literacy, knowledge, and awareness.

Elmustee is about water conservation and recycling.

Elmustee is about energy self-sustenance.

Elmustee is about food self-reliance.

Elmustee is about preservation of local languages and cultures.

Elmustee is about continuing education.

Elmustee is about social justice.

Elmustee is about gender equality.

Elmustee is about accessibility provided to all differentlyabled people.

Elmustee is about the simple idea that no matter where one lives in the world, one has a right to enjoy the privileges of a first world society.

ABOUT THE ELMUSTEE READING MATERIAL

Three books describe the concept, the technologies to be used, and the workings of Elmustee, a membership community. *Elmustee, A Growing Island of Hope*, taking the form of a novel, helps readers understand the evolution of ideas behind Elmustee. *Elmustee, The Membership Experience*, touches upon the issues applicants wanting to join Elmustee must be aware of, before being inducted into the community. This book, *The Elmustee Technologies,* describes the technologies that will be used at the membership community.

www.ingramcontent.com/pod-product-compliance
Lightning Source LLC
Chambersburg PA
CBHW071435180526
45170CB00001B/349